FUNGAL BIOLOGY:

Understanding the fungal lifestyle

FUNGAL BIOLOGY:

Understanding the fungal lifestyle

D.H. Jennings

Department of Genetics and Microbiology, University of Liverpool, Liverpool, UK

G. Lysek

Institute of Systematic Botany and Plant Geography, Free University of Berlin, D-14195 Berlin, Germany

βIOS
SCIENTIFIC
PUBLISHERS

A CIP catalogue record for this book is available from the British Library.

ISBN 1 85996 150 9

BIOS Scientific Publishers Ltd
9 Newtec Place, Magdalen Road, Oxford OX4 1RE, UK
Tel. +44 (0) 1865 726286. Fax +44 (0) 1865 246823
World-Wide Web home page: http://www.Bookshop.co.uk/BIOS/

DISTRIBUTORS

Australia and New Zealand
    DA Information Services
    648 Whitehorse Road, Mitcham
    Victoria 3132

India
    Viva Books Private Limited
    4325/3 Ansari Road
    Daryaganj
    New Delhi 110002

Singapore and South East Asia
    Toppan Company (S) PTE Ltd
    38 Liu Fang Road, Jurong
    Singapore 2262

USA and Canada
    Books International Inc.
    PO Box 605, Herndon, VA 22070

The cover illustration of *Boletus luridus* is reproduced from the original watercolour by E. Ludwig with kind permission, and will be published in *Pilzokompendium*, IHW-Verlag, Eching.

Typeset by Chandos Electronic Publishing, Stanton Harcourt, UK.
Printed by Biddles Ltd, Guildford, UK.

Contents

PART 2: THE ENVIRONMENT

5. Water: living with desiccation   67

6. Oxygen and lack of it   75

7. Using light   81

8. Withstanding extremes of temperature   89

Appendices

Abbreviations

ADP	adenosine diphosphate
ATP	adenosine triphosphate
NAD	nicotinamide adenine dinucleotide
NADP	nicotinamide adenine dinucleotide phosphate
V-A	vesicular–arbuscular
VLP	virus-like particles

Glossary

Table 11.1 (p. 108) should be consulted with respect to terms describing the mating and nuclear relationships of fungi.

Absorption zone: region behind the hyphal tip where the bulk of the nutrients is absorbed.

Aeciospore: first dikaryotic spore in the life cycle of rust fungi, often infecting the second host.

Agar: jelly produced by dissolving (usually 1% w/v) in hot water (or an appropriate solution of nutrients, when it is termed nutrient agar) an extract of the cell walls of certain red algae. Few fungi can break down the carbohydrate forming the gel matrix (a complex of galactans, some of which are sulphated), so nutrient agar provides an admirable solid surface for mycelial growth. The low percentage of solid in agar means that it does not have much effect on water availability.

(An)aerobic: conditions with (without) oxygen.

Anamorph: the term for the form of reproduction, in this case vegetative (giving the conidial or imperfect state), and also the taxonomic name of the species with this form of reproduction.

Anastomosis (plural **anastomoses**): the fusion of one hypha with another such that there is protoplasmic connection between them. The connection between two radially growing hyphae may be through the branch of one making the fusion and thus forming a bridge.

Apical growth zone: the growing (extending) hyphal tip and the adjacent regions of the hypha that provide the metabolites and enzymes necessary for growth.

Aplanospore: spore without a flagellum and thus non-motile.

Appressorium (plural **appressoria**): a hyphal branch that fixes the mycelium to the surface of the host; it often will be the starting point for invasion of the underlying cells.

Ascus: a sac-like cell in which ascospores are produced as free cells after karyogamy and meiosis (and often after an additional mitosis). *See also* Karyogamy.

Autolysis: self-digestion of senescent hyphae.

Axenic: free from undesirable microorganisms; uncontaminated.

Balanced growth: growth in which the bulk composition of the mycelium remains relatively constant over a significant period of time; dry matter increase is exponential during the phase of balanced growth. It occurs under conditions of nutrient sufficiency.

Basidium (plural **basidia**): structure produced by basidiomycetes in which karyogamy and meiosis and on which spore formation occurs. *See also* Karyogamy.

Biotrophic: growing by the exploitation of living organisms.

Blastic: formation of conidia by the swelling of hyphae *before* the spore initial is delimited by the formation of a septum (*Figure 12.3,* p. 121). *See also* Thallic.

Brown rot: fungal decay of wood in which virtually only the cellulose is degraded by the fungus; there is relatively little change to the structure of the lignin.

Nevertheless there is sufficient chemical change for the residue to turn brown. *See also* White rot.

Chlamydospore: resting spores formed from parts of vegetative hyphae.

Clamp connection: narrow hyphal bridge containing a septum which leaves the main hyphal compartment close to a septum and joins the next compartment on the other side, again close to the septum (see Life Cycle 5, p. 142).

Colony: either the coherent mycelium or a mass of cells (such as those of a yeast) of one origin. *See also* Mycelium.

Composting: degradation of organic material, predominantly of vegetable origin, of sufficient bulk to allow the build-up of heat, such that the process of decomposition is speeded up (see p. 62).

Conidium (plural **conidia**): exogenously produced spore (not produced by cytoplasmic cleavage or free-cell formation), which is found exclusively in and is typical of fungi.

Coprophilous: growing in dung.

Cryophilic: able to live at low temperatures.

Dermatophyte: a fungus living on and/or infecting human or other animal skin.

Dermatomycoses: fungal diseases of the skin.

Dispersal: transport of spores to other sites.

Endophyte: a fungus living within plants, often without causing visible symptoms.

Exoenzyme: an enzyme released by a fungus, which breaks down molecules external to the cell or hypha.

Fairy ring: a ring, which will be one of a sequence, of fruiting organs, generally of basidiomycetes and most frequently in grassland.

Fruiting organ (fruit body): complex structure built up from hyphae in or on which meiospores are produced. *See also* Meiospore.

Fungi imperfecti (or **imperfect fungi**): fungi without a (known) sexual state; taxonomic name **deuteromycetes.**

Gemma (plural **gemmae**): large resting spore of oo- and zygomycetes.

Germ tube: that hyphal tip that breaks through the spore wall and if the fungus is septate, before the first septum is formed.

Habitat: environment of a distinct type. *See also* Substrate.

Haustorium (plural **haustoria**): short hyphal branches that invade a cell (and therein can branch further) and absorb the contents as they are made soluble (see *Figure 3.4,* p. 37).

Helicospores: helical- or spiral-shaped spores.

Hook: the tip of an ascogenous (dikaryotic) hypha, which is bent to form a hook (see Life Cycle 3, p. 138). This process of bending is the initial step in the formation of an ascus and appears to be the means by which both nuclei divide simultaneously.

Host: any living organism living together with or invaded and infected by another (parasitic) organism.

Hypha (adjective **hyphal**; plural **hyphae**): the individual unit of the filamentous growth form.

Idiophase: phase of unbalanced growth when secondary metabolites and reproductive structures may be formed, mostly after the end of trophophase. *See also* Trophophase.

Karyogamy: fusion of two nuclei.

Lichen: the self-supporting association of a fungus (**mycobiont**, almost always an ascomycete fungus) and an alga or cyanobacterium (often **phycobiont** but more correctly **photobiont**).

Lignicolous: living on wood and using it as a substrate.

Meiospore: spore formed after meiosis.

Mitospore: spore containing nuclei of mitotic origin.

Mushroom: a member of the Agaricales or Boletales (see p. 147), and also some ascomycetes such as morels (*Morchella*), with an edible fruit body. *See also* Fruiting organ.

Mycelium (plural **mycelia**): network of hyphae.

Mycophilous: growing by exploitation of fungal mycelia or fruit bodies.

Mycorrhiza (plural **mycorrhizas**): a fungus living mutualistically with the roots of a higher plant (see *Figure 3.7*, p. 41). That where the bulk of the mycelium is on the outside of the root is known as an *ectomycorrhiza;* that where the bulk is inside an *endomycorrhiza*.

Necrotrophic: growing by first killing (the cell of) the host organism or mycelium.

Parasexuality: those mechanisms that achieve alteration of the genome without mutations, meiosis or fertilization.

Pellet: those fungal colonies of spherical or semi-spherical shape produced in agitated liquid culture inoculated with spores or small hyphal fragments.

Petri dish: transparent plate (glass or plastic) with vertical sides and overlapping lid in which microorganisms are cultured, usually on nutrient agar. *See also* Agar.

Phialide: a hyphal branch on which a basipetal succession of conidia develop without increase in the length of the phialide itself.

Phylloplane: surface of a leaf together with the contiguous environment and the organisms living therein.

Phytopathogenic: bringing about plant disease.

Planospores: spores carrying flagella and hence capable of active movement in a liquid environment.

Plasmogamy: fusion between two sexual cells.

Predacious: living by the exploitation of captured and killed small animals.

Psychrophilic: showing optimal growth at low temperatures.

Reproduction: any process leading to daughter colonies and/or daughter nuclei. *See also* Colony.

Reserve substance: material (usually polymeric) stored for later use in metabolism.

Resource: material external to the fungus (or other organism) that can be used for growth.

Rhizomorph: aggregation of hyphae into a linear organ capable of translocating water and nutrients over long distances.

Rhythmic growth: formation of concentric bands at regular intervals in a mycelium; often leading to rhythmic formation of reproductive structures.

Rust fungus: fungus belonging to the Uredinales.

Saprotrophic: living on dead organic material.

Sclerotium (plural **sclerotia**): a complex permanent structure, with a hard pigmented outer covering, built from densely packed hyphae, which, under appropriate conditions, can 'germinate' to give a normal vegetative mycelium.

Senescence: ageing of hyphae often accompanied by autolysis. *See also* Autolysis.

Septum (plural **septa**): cross-division of hyphae with a central pore (not to be confused with a cross-wall where no pore is present).

Sink: structure (or organism) that draws on nutrients for metabolism/storage from another part of the same organism (or from another organism) and in this way may alter the pattern of nutrient flow within the system.

Smut fungus: fungus belonging to the Ustilaginomycetes (also Ustomycetes).

Soft rot: fungal decay of wood via the formation of cavities in wood cell walls, leading to a softening of the wood and increased water retention within it.

Spermatium (plural **spermatia**): small spore (conidium) which acts only in the transference of nuclei.

Sporangiophore: the structure supporting the sporangium.

Sporangium (plural **sporangia**): structure containing spores.

Spore: any distinct part of a mycelium capable of bringing about dispersal.

Staling: exogenously, or endogenously, caused cessation of hyphal/mycelial extension.

Storage: accumulation of nutrients within one part of a cell/hypha often in the form of polymeric material.

Substrate: the space in which a fungus grows and the material filling this space and thus providing the nutrients for the fungus. *See also* Habitat.

Succession: regular sequence of fungi using different material within a substrate.

Symbiosis: mutualistic combination of two (or more) organisms.

Teleomorph: the term for the form of reproduction, in this case sexual, and also the taxonomic state of the species with this form of reproduction.

Thallic: method of formation of conidia in which there is swelling of the hypha after the conidial initial has been delimited by one or more septa (see *Figure 12.3;* p. 121). *See also* Blastic.

Thermophilic: lives at high temperatures.

Toadstool: a member of the Agaricales or Boletales (see p. 147) with an inedible fruiting body. *See also* Fruiting organ.

Translocation: movement of water/nutrients over long distances within hyphae/mycelia.

Trophophase: phase of extension, nutrient absorption and balanced growth. *See also* Balanced growth.

Uredospore: dikaryotic spore of rust fungi produced in the second host and capable of reinfecting it.

Vector: factor bringing about the transport of organisms, spores or other reproductive structures

Vegetative state: mycelium in which there are no reproductive structures.

White rot: fungal decay of wood in which both cellulose and lignin are broken down (though not necessarily in equal amounts); the residue is white and fibrous. *See also* Brown rot.

Yeast: a fungus which forms distinct cells; reproduction by budding or division.

Zoospores: motile spores. *See also* Planospores.

Preface

This book has been written for those starting a course of mycology, whether it be one devoted solely to the fungi or, as seems more likely, a course on microbiology in which the study of fungi has a significant role. There is also another target group: there are many now, particularly in the field of molecular biology, who are undertaking research on fungi and who have never received formal training in mycology. This book is also for such persons. We hope it will allow them to see their research in a wider perspective.

The text is focused on the general biology of fungi, because we believe that it is this approach to mycology that provides the best foundation for an understanding of the fungi. Consequently, we have tended to shun concentrating on specific species. So, unlike many mycological texts, there is very little detail about life cycles and about details of reproduction. Nevertheless, in Appendix A we have provided details of five representative life cycles for what we believe are amongst the more significant fungi. We advise readers to consult other texts for details of morphology and reproduction of specific fungi. Details of such texts are given in the Further reading list in Appendix C.

Mycology can be daunting because of the considerable terminology used by those studying fungi. We commend those reading this book, who know little about mycology, to make full use of the glossary at the front of the book.

We wish to thank Mr H. Lünser for producing the many fine illustrations. Inspection of them indicates only part of his skill; knowledge of the suggested outline drawings by ourselves increases one's admiration of his ability to conjure up the finished product. We are also grateful to those who gave us permission to use diagrams and tables, and particularly those who provided photographic prints. Due acknowledgement is given at the appropriate place in the text.

We are especially grateful to Professor Birgit Nordbring-Hertz and Dr Susan Isaac, who read a draft of the text, for their expert opinion and advice. Any errors must be laid firmly at our door.

David Jennings
Gernot Lysek

Foreword

This book originates from a series of invited lectures* given by the authors in Berlin and Liverpool respectively, and aimed primarily at first and second year students of biology. These lectures have now been broadened and compiled into a text on fungal biology, covering a wide range of fungi in relation to their typical environments, and presenting a comprehensive body of information that will enable students to grasp the basic mechanisms involved.

The authors present a fascinating picture of fungal physiology seen from an environmental perspective. Basic knowledge is presented about hyphal and colony growth as well as differentiation and reproduction, all in relation to environmental factors such as nutrients, water, oxygen, light and temperature. The text is accompanied by a large number of illustrations, both reference material and new, instructive drawings and micrographs, which greatly increase its readability.

I was privileged to be able to read an earlier version of this book and was greatly impressed by its interesting ecophysiological approach. The authors are to be congratulated on the compilation of such a wealth of physiological, ecological and experimental mycology.

Birgit Nordbring-Hertz
University of Lund, Sweden

* Supported by the Erasmus scheme of the European Community.

Introduction

If one had to design an organism that could digest and (re-)mineralize natural organic substances (be they solid, liquid or gaseous, be they dead or part of a living organism), do it with high efficiency and essentially without temporal and spatial limits and also possess flexibility to adapt to new substances, one would end up with a mycelial fungus. These organisms are, if taken as a whole, omnivorous, in that they can consume a huge variety of organic compounds.

Since they can only be stopped by the exhaustion of the substrate, if that substrate is accessible, fungi are capable of limitless growth. This is demonstrated by the recent evidence that the mycelium of a fungus (*Armillaria bulbosa* growing in a forest soil) is capable of yielding amongst the largest biomass of any single living organism. Furthermore their hyphae can transport nutrients accumulated from rich sources to sites where the same nutrients are in short supply such that the hyphae can continue to extend or differentiate into reproductive structures. Fungi even have the means to continue nutrient digestion and absorption during reproduction. Though other factors like dryness, low or high temperature, antifungal agents can interrupt growth, fungi are often remarkably tolerant of these external factors. Not surprisingly, therefore, fungi can be considered all-invasive in terms of space, because they can colonize natural organic substrates on high mountains as well as in the sea and in even more salty environments, in heavy metal-contaminated soils as well as in the dust on remote bare surfaces, living organisms as well as non-living remains. Fungi are virtually everywhere; as the result of their very effective means of reproduction and spore dispersal, fungi are always present when a suitable substrate becomes available.

The properties of fungi (other than yeasts) reside in the activities of their growing mycelia, or, at the microscopic level, in the activities of the hyphae, which are long filaments consisting of a tube of polysaccharidic material, mainly β-linked polymers of glucose (glucans, including cellulose) and of *N*-acetyl glucosamine (chitin), the actual mixture depending on the fungal group (*Table 1*). This tube, the wall, encloses the cytoplasm with its outer boundary the plasmalemma and within the cytoplasm, the nuclei and other organelles, including the vacuolar system and its bounding membrane, the tonoplast.

The solutes in the cytoplasm and vacuolar system create the osmotic potential (expressing itself as a hydrostatic pressure) that generates the turgor which is almost balanced by the inwardly directed resistance to stretching of the hyphal wall. The small positive hydrostatic pressure generated brings about hyphal extension. Thus far hyphae seem to resemble the cells of green plants, but there are large differences.

Thus there are no cross walls to divide a hyphal filament. There can be partial divisions in the form of septa (singular=septum). Septa appear at first sight like a cross wall but they possess pores, allowing communication longitudinally in the hypha in the form of the movement of water and solutes as well as organelles and even nuclei. In most of what used to be called formally 'phycomycetes' septa are absent (*Table 1*), thus hyphae are unique among wall cells for the continuity of their protoplasm.

Further differences include the absence in fungi of the characteristic organelles of green or photosynthesizing plants, the plastids. Therefore fungi depend completely on organic nutrients for the combined carbon necessary for growth. On the other hand, fungi can adjust the internal osmotic potential to values that far exceed the abilities of green plants.

Fungi are organisms of decay, but they are not alone in this activity; at the microscopic level, bacteria and protozoa act similarly. However, the environments in which all these decay organisms live are not homogeneous. Fungi, because of their hyphae, can spread from one patch of nutrients to another, using the energy in the nutrients in the first patch for the growth of the hyphae to the next patch. Also, because the hyphae can aggregate into organs, fungi are able to colonize other nutrient sources considerable distances away, either by root-like organs (rhizomorphs) or by spores released into the turbulent air above the boundary layer next to the soil, a process aided considerably by the production of spores in appropriately tall fruiting bodies such as mushrooms.

The study of the fungi, mycology, sits uneasily between microbiology and botany. Many of the techniques used to study the prokaryotic bacteria are applicable to the eukaryotic fungi, but fungi have walls of a similar kind to those of green plants and produce complex structures akin to those in the green plant world. So fungi cannot be considered as filamentous bacteria, nor are fungi an adjunct to the green plant world. Fungi are unique organisms. This book is about them, providing an overview of their fundamental properties and their activities in the natural world. *Table 1* summarizes the most important characters of the major classes of the Fungal Kingdom.

Table 1: The main classes of the fungal Kingdom and their most significant morphological and biochemical features

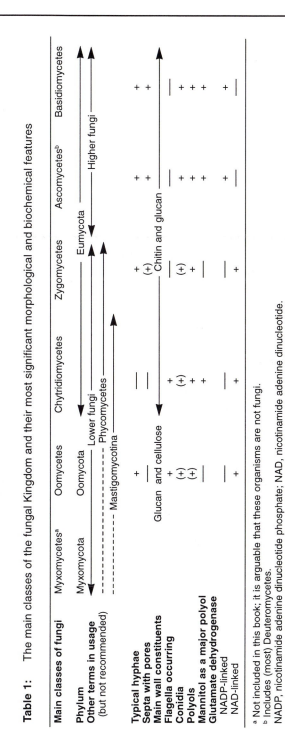

Main classes of fungi	Myxomycetes[a]	Oomycetes	Chytridiomycetes	Zygomycetes	Ascomycetes[b]	Basidiomycetes
Phylum	Myxomycota	Oomycota	← Eumycota →			
Other terms in usage (but not recommended)	← Lower fungi →	← Phycomycetes →			← Higher fungi →	
	← Mastigomycotina →					
Typical hyphae	—	+	—	+	+	+
Septa with pores	—	—	—	(+)	+	+
Main wall constituents		← Glucan and cellulose →		← Chitin and glucan →		
Conidia	—	(+)	+	(+)	+	+
Polyols	—	(+)	+	+	+	+
Mannitol as a major polyol	—	—	—	—	+	+
Glutamate dehydrogenase						
NADP-linked	—	—	—	—	+	+
NAD-linked	+	+	+	+	—	—

[a] Not included in this book; it is arguable that these organisms are not fungi.
[b] Includes (most) Deuteromycetes.
NADP, nicotinamide adenine dinucleotide phosphate; NAD, nicotinamide adenine dinucleotide.

Chapter 1

Hyphae and hyphal extension

1.1 Introduction

Fungal hyphae as they grow and branch produce a network of filaments which constitutes the fungal mycelium (*Figure 1.1*). The mycelium expands by extension of individual hyphae which exhibit polar growth, that is they grow at their tips. Expansion of the mycelium is limitless if hyphae can continue to extend on the surface from which they gain their nutrients. Nevertheless, changes take place as the mycelium ages and as that part of the substrate on which it is growing is no longer able to provide nutrients.

1.2 The fungal mycelium

If one moves from the growing margin of a mycelium along a longitudinal axis, four different zones can be found which correspond to different ages or developmental stages.

- The apical growth zone with the extending hyphal tip and the adjacent part that provides the material and organelles required for hyphal elongation.
- The absorption zone where there is uptake of nutrients (this zone partially overlaps with the apical growth zone).
- The storage zone in which a proportion of the absorbed nutrients are stored as reserve substances.
- The senescence zone, the oldest part of the mycelium, is characterized by the presence of dark pigments and lysis may eventually take place.

1.3 The apical growth zone

The apical growth zone contains the very tip of the hypha, which is in the form of a dome (*Figure 1.2*). Here the hypha increases the area of its wall. This increase is

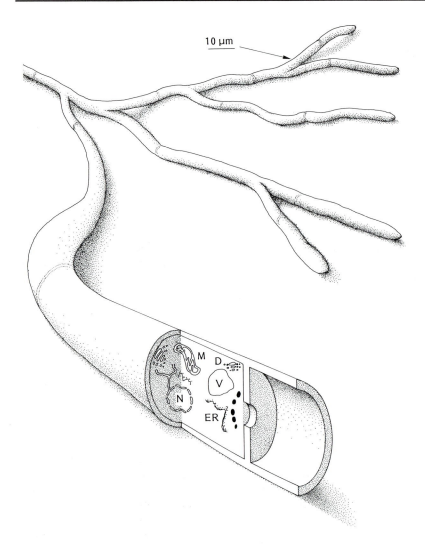

Figure 1.1:   The hyphal branching system. One branch (a hypha) is sectioned to show the septum with the pore and some characteristic features of the protoplasm. N, nucleus; ER, endoplasmic reticulum; D, dictyosome or Golgi apparatus; V, vacuole; M, mitochondrion. The dark bodies are Woronin bodies, which are composed of protein and which can block the septal pore when a hyphal compartment becomes physically ruptured, such that the contents of undamaged compartments are not lost externally.

achieved by insertion of precursors, such that microfibrils made up of aggregations of unbranched long-chain polymers [in the eumycota the predominant polymer is chitin; in the oomycetes there is β-(1-3),β-(1-6)-glucan and cellulose] are increased in length

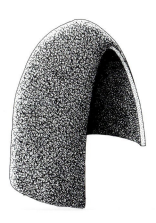

Figure 1.2: The dome-shaped wall of the hyphal tip (part cut away), showing the general organization of the fibrillar material in the wall.

and smaller molecular weight polymers (such as xylo- or galacto-mannoproteins or glucan) intercalated between the fibrils. The incorporation of new material extends the wall from the dome in a cylindrical manner.

To fuel this incorporation, there is a permanent flow of material into the hyphal tip. This material includes not only the above-mentioned precursors but also the enzymes necessary for synthesizing the wall material. There is still argument as to whether hydrolytic enzymes are also required to break bonds to allow new material to be inserted into the wall. There is good evidence that, at the tip, the microfibrils can slide apart under turgor and new material can be intercalated which only becomes firmly bound to the microfibrils after a period of time, when the wall is starting to take up its cylindrical shape. Irrespective of that argument, there is much evidence to suggest that precursors and enzymes are enveloped within a lipid membrane and transported in this form as a vesicle to the tip. Here the vesicle membrane fuses with the plasmalemma on contact with it, such that the vesicle is opened with its inside seemingly becoming the outside surface of the plasmalemma. In this way the contents of the vesicle come into direct contact with the expanding wall. The vesicles are formed in the rear part of the apical growth zone and are presumed to originate in the dictyosomes, the same organelles that produce excretory and outer covering materials in green plants. The vesicles, together with mitochondria and endoplasmic reticulum, fill the region just behind the dome but nuclei are absent.

This raises the question as to how polar growth is maintained. There is good evidence that the activities of the next zone, the absorption zone (see Section 1.4), appear to be important (*Figure 1.3*). This region extrudes H$^+$, protons, at the expense of metabolic energy by means of what is known as the 'proton extrusion pump'. The net negative charge inside the hypha results in the absorption of positive ions, particularly potassium. The potassium ions thus accumulated in all probability diffuse to the tip. Likewise, protons, which are known to move in the reverse direction as a consequence of entering the tip, which is known to be permeable to the ion (along with amino acids in a few instances and probably phosphate, which both can enter the tip from the

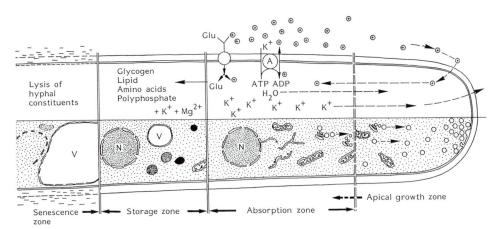

Figure 1.3: The main ultrastructural features (lower half) and main processes and storage compounds in the storage zone (upper half) within the principal zones of a growing hypha. A, ATPase; Glu, glucose; N, nucleus; V, vacuole; →, active solute movement; ⇢, solute movement by diffusion or through water flow; ⊕ hydrogen ions/protons. The other organelles, endoplasmic reticulum, dictyosomes, mitochondria and vesicles, have not been labelled but are readily recognized.

external medium). Indeed, there is a circulation of protons through the proton extrusion pump, then through the medium to the tip, whence there is a return to the cytoplasm via proteins transporting the protons across the tip membrane. This circulation of protons, because it carries charge, can be detected as an electric current, the circuit being the path of proton movement as just described. There could be a circulation of potassium in the reverse direction but this has yet to be demonstrated. The gradient of protons (and the reverse gradient of potassium?) may have a significant role in guiding vesicles to the hyphal apex.

At the apex there is a marked aggregation of vesicles to a density which is visible by the light microscope as a discrete organelle called the 'Spitzenkörper' by its discoverer

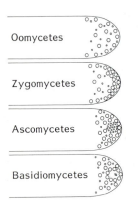

Figure 1.4: Diagram showing the larger vesicles and their arrangement in extending hyphae of (from top down) oo-, zygo-, asco- and basidiomycetes. The Spitzenkörper is the mass of smaller vesicles shown at the apex of the ascomycete hypha.

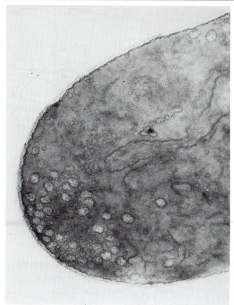

Figure 1.5:   Electron micrograph of the growing hyphal tip of *Aspergillus ochraceous* showing the aggregation of vesicles with the Spitzenkörper in the centre. Photograph by R. Dargent.

Girbardt (*Figures 1.4* and *1.5*). He also found that the Spitzenkörper is only found in growing hyphae; when hyphae cease to elongate, it is lost. There is evidence that the Spitzenkörper functions as a final distribution centre for wall-destined vesicles.

It is not clear how the vesicles reach the tip. There is within the hypha a complex cytoskeleton made up (in the simplest terms) of microfilaments and microtubules (*Figure 1.6*). The former are made of the contractile protein, actin. Thus if vesicles become attached to the microfilaments, the contractile action of the latter could lead to vesicle movement to the tip.

1.4 The absorption zone

The steady production of vesicles with their various contents and their directed movement to the tip from some distance behind it, is dependent on the supply of nutrients from the external medium across the plasmalemma in the absorption zone. The movement of solutes across the membrane is dominated by the activity of the proton extrusion pump referred to earlier. This pump is an enzyme, an ATPase which uses the free energy of the terminal pyrophosphate bond of ATP to bring about the extrusion of protons. As we have indicated, this loss of protons leads to an accumulation of potassium within the hypha.

Thus the outside of the hypha is acidified by the continuing excretion of protons. The difference between the inside and the outside may reach 1000-fold. This means a pH difference of 3.0 pH values between the medium and the cytoplasm, and, since this

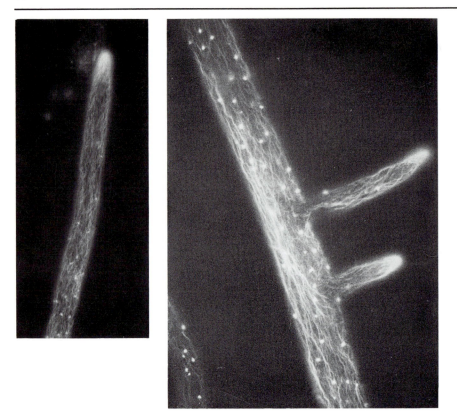

Figure 1.6:   The actin component of the cytoskeleton of hyphae of *Saprolegnia ferax* made visible by fluorescence microscopy using rhodamine-phalloidin staining. Photographs by I.B. Heath.

difference of hydrogen ions is only a part of the electrochemical potential gradient generated across the plasmalemma, it also means that the cytoplasm can be 200–300 mV more negative than the external medium. It is the free energy residing in the proton electrochemical gradient that drives the movement of other solutes across the plasmalemma.

The general mechanism by which this solute transport is brought about is as follows. Protons external to the plasmalemma, because of their lower concentration inside the hypha and the more negative cytoplasm, if given the opportunity, will diffuse into it. This diffusion gradient, driven by the electrochemical potential gradient across the membrane, will itself drive the movement of other solutes across it, bringing about the accumulation within the hypha of, for example, potassium, referred to above, and organic solutes such as glucose and amino acids. Mechanistically, the diffusion of protons into the hypha is mediated by permeases/carriers, proteins which by binding protons and the accompanying solute can move them both across the otherwise impermeable membrane. This specific type of transport of nutrients into the cytoplasm driven by the proton diffusion gradient is called 'proton symport' (*Figure*

1.7). Table 1.1 gives a list of substances that have been shown to be accumulated by such proton symports and the fungi that have been used in relevant studies. Such nutrients consumed from the environment, when in the cytoplasm, will be involved in supporting the ongoing formation of vesicles that keep running the extension of the tip. But these nutrients, particularly potassium, also contribute to the osmotic potential of the hypha and therefore to the osmotic pressure within it. As indicated above, the osmotic pressure generates the turgor for growth, consequently regulation of the potassium content of a hypha will regulate its cytoplasmic osmotic value and therefore its turgor. In physiological terms, as is the case with all walled cells, the growth of a hypha requires that the protoplasmic water potential be more negative than that of the medium.

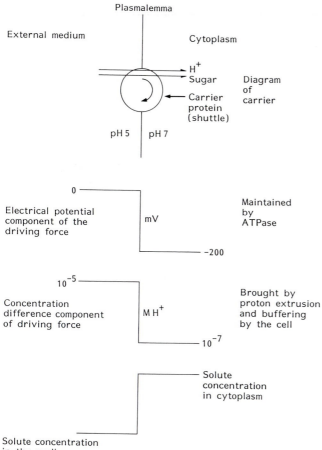

Figure 1.7: The membrane processes leading to the accumulation of nutrients (in this particular instance glucose) in a fungal hypha.

Table 1.1: Nutrients that have been shown to enter a fungus via a proton symport, that is, the nutrients are co-transported with protons

Nutrient	Species
Potassium	*Neurospora crassa*
Glucose	*Dendryphiella salina* *Neurospora crassa* Many yeasts (but not *Saccharomyces cerevisiae*)
Amino acids	*Achlya bisexualis* *Neurospora crassa* *Penicillium* spp.
Phosphate	*Saccharomyces cerevisiae*

There is a further interaction between the hyphal apex and the absorption zone. At the apex, enzymes are released into the external environment which are capable of attacking substances that otherwise would be unavailable to the fungus either due to their insolubility or their molecular size which prevents movement across the plasmalemma. These enzymes, some of which may also be involved in hydrolysing bonds in the process of hyphal extension, are termed exoenzymes or extrahyphal (often extracellular). The end-product of this process of 'digestion' is the production of molecules which can be taken into the cytoplasm of the hypha, via the permeases of the absorption zone. It is the armoury of extracellular enzymes produced by fungi that makes them so effective in breaking down the wide range of insoluble material found in nature, for example, wood, leather, hair, crude oil. This combined action of the tip, which extends and excretes extrahyphal enzymes, and the absorption zone, where the products of enzyme action are transported into the cytoplasm, is unique to fungi and is referred to as 'trophic growth' and the period in which it occurs as the 'trophophase'. Parenthetically, there are circumstances when hyphae are involved in non-trophic growth as when they grow into the air.

Trophic growth is not only a very effective means of using the environmental resources but also the best way to capture nutrients in the face of other microorganisms, bacterial and animal, capable of using the same substrates. Fungi can, by their growth habit, more readily move over the substrate, capturing it from possible competition. However, in nature, the manner of that growth will depend on whether or not the fungus is in an environment in which the colony finds it hard to obtain the necessary nutrients. If they are scarce, the hyphae are sparse and thin with large distances between the branches. This form has been called 'hunger mycelium' (*Figure 1.8*). It would seem that all the available nutrients and energy are used for the elongation of very few hyphae, in which the contents, cytoplasm and organelles, are continuously being incorporated into the growing end of the hyphae in order to make full use of the available resources. The rate of extension is not necessarily reduced. This form of hyphal growth can be seen as an effective means of searching for new substrates.

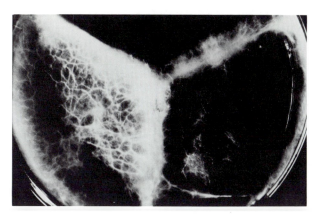

Figure 1.8: Regulation of hyphal density through nutrient availability; mycelium of *Serpula lacrymans* growing (in a glass dish divided equally into three compartments separated by partitions) over nutrient-rich (left compartment) and nutrient-poor agar (in the two compartments to the right).

However, an environment with a good supply of nutrients will allow a high rate of metabolism. Also, one can conceive, there could be a high rate of vesicle formation, more than required for continuous apical growth. The excess of vesicles over those destined for transport to the apex might eventually come into contact with the plasmalemma near their origin (*Figure 1.9*). Here they fuse with the membrane, their

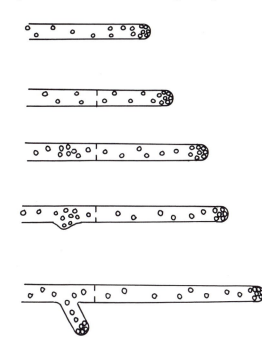

Figure 1.9: Diagram showing the role of septa and vesicle accumulation in the formation of a hyphal branch. The density of circles indicates the density of vesicles.

contents being released into the hyphal wall, copying events at the hyphal tip. Now all other functions of an apex commence. There is certainly entry of protons; indeed Harold has said that "where protons enter [a hypha] there is a tip". In this way, an apex is created, which starts elongating as a new and independently growing branch. One can see how the number of branches, and also the density of the resulting mycelium, might be adjusted to the exogenous nutrient supply. Thus, in nutrient-rich habitats dense mycelia are formed, whereas in nutrient-poor habitats, as described above, there are only sparse mycelia, as a consequence of the supply of vesicles being only sufficient to maintain the extension of the leading hypha. There are some chemical compounds (paramorphogens), a number of which are analogues of glucose, that cause an extremely high density of branching. Colony expansion occurs at a very low rate, giving the impression that growth is inhibited, yet dry matter production is hardly affected. The cause of the phenomenon is not clear.

If we look at the growth of individual hyphae, they extend at a constant *linear* rate. On the other hand, the number of hyphal apices produced by the expanding colony increases at an *exponential* rate. It is thus the formation of branches in a fungal colony, that is the formation of new hyphal apices, which is the equivalent to the production of a new cell in a yeast or bacterial colony.

1.5 The storage zone

It is likely that, in nature, a fungus will be in a habitat where there are sufficient nutrients for growth. Indeed, some nutrients may be in excess of those required for balanced growth. Nevertheless, they may still be absorbed into the cytoplasm, there to be converted into an insoluble form or into a form sequestered in the vacuolar system. In this way nutrients are stored as reserve material. Carbon is stored as glycogen or lipid; nitrogen as amino acids in the vacuolar system or as protein, while phosphorus is stored in condensed form as polyphosphate, often in the vacuole, where, through its negative charges, it binds potassium, magnesium or positively charged amino acids such as arginine. Storage in insoluble form means minimum effect osmotically and, indeed, the conversion of metabolites into insoluble compounds and vice versa is one way in which a fungus can respond to osmotic change.

The reserve substances are stored until the exogenous supply of their precursors ceases. The reserve substances can sometimes be utilized for further hyphal extension. However, the activation of reserves usually marks the start of processes other than trophic elongation. The mycelium leaves the trophophase and enters the 'idiophase', in which there is only a low rate of growth, if any; the phase being distinguished by differentiation, often involving reproduction, taking place within the mycelium. It is during this stage that reserves are activated and the soluble products incorporated, as either soluble or insoluble reserve material, into the reproductive propagules produced. Large and thick-walled spores (chlamydospores) or sclerotia contain very significant reserves, which are activated and consumed during germination, until the newly formed colony has developed sufficiently for its own trophic extension.

1.6 The senescence zone

After reproduction, that is, in the still older parts of the colony, the mycelium starts to senesce. Often the first visible sign is the formation of greenish or black pigments. In some fungi, pigmentation starts soon after the exploitation of the substrate. Nevertheless pigmentation is an indication of hyphal ageing.

Later the mycelium disintegrates (*Figure 1.10*). This process is mainly autolytic; many fungi contain enzymes which at the end of the lifetime of the mycelium act as self-destruction equipment. In growing mycelium, these enzymes are confined to a compartment of the vacuolar system, from which they are released when intermediary metabolism runs down, or after a hypha is damaged physically.

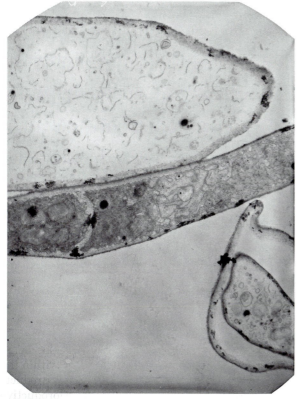

Figure 1.10: Electron micrograph of a mycelial pellet of *Dendryphiella salina* which has been growing in liquid culture until the nutrients are almost exhausted; both living (middle) and senescent (top and bottom) hyphae can be seen.

Senescence, the last phase of mycelial development, is often accelerated, or accompanied, by alterations to its DNA, which contains senescent parts, termed 'virus-like particles' (VLP), consisting of rings of fungal DNA. These VLPs have been found in some fungi that occur naturally on substrates that are limited in size. *Podospora anserina*, a species that shows very markedly this form of senescence, lives in dung balls. In the laboratory, a culture grows only for a limited time and distance and then turns senescent. This regularly occurring senescence might be the means by which the fungus restricts its mycelial extension to the range commensurate with the volume of nutrient resource in nature, such that all the nutrients are exploited and reproductive propagation carried out, yet vegetative growth is terminated in an efficient manner.

The developmental sequence, with its different phases outlined in this chapter is the basic model. Within the Fungal Kingdom this model is altered and adapted to meet environmental and developmental demands. It is a fair assumption to say that there are as many adaptations as there are fungal species.

In addition to expansion by hyphae, some fungi grow and propagate as individual cells called yeasts. For this widespread, but nevertheless special form, see Chapter 6, p. 78.

Chapter 2

The mycelium

2.1 Introduction

As will be shown below, the growth of the ideal mycelium, that is one which is growing in a homogeneous environment, is readily described. Such a description shows how hyphae fill the space they invade and how growing hyphae interact with their environment, with particular regard to the supply of oxygen. Under natural conditions, where nutrients are only patchily distributed, the situation is more complicated. However, it is clear that, under such circumstances, the uniqueness of the mycelial form as an interconnected system can overcome the heterogeneity of the environment by shifting nutrients through the hyphae, thus allowing them to cross those areas in which nutrients are lacking.

2.2 The ideal mycelium

As shown in Chapter 1, hyphae grow at their tip in a potentially unlimited manner. In a homogeneous environment, this growth and accompanying branching results in an expanding spherical colony. *Figure 2.1* shows that, after only a small number of branchings, a radial colony is formed. Thus, if the angle of branching is 60°, four branchings are sufficient to produce at least one hypha extending in the direction opposite to that of the germ tube; if the angle is 30°, six branchings are needed. In a very short time all the growing hyphae are forming the surface of a sphere (*Figure 2.1e*), which expands at the rate of hyphal extension in a potentially unlimited manner.

Inside the sphere, the mycelium consists mainly of radially running hyphae and hyphal branches, which fill the space at an equal density, ensuring maximal exploitation of the available nutrients. In this system, anastomoses are formed as short hyphal bridges between neighbouring hyphae. These anastomoses thus form lateral interconnections, through which nutrients, water or even organelles can be shifted laterally. The role of these interconnections is seen in *Figure 2.2*, where, in a

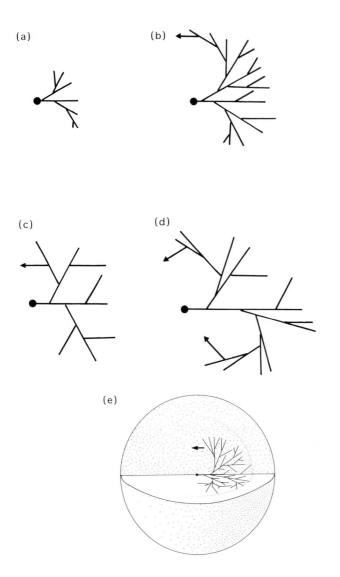

Figure 2.1:   Diagram to show how a spherical colony develops after germination of a spore, such that, independently of the angle and frequency of branching, a spherical colony develops. (a), (b) Branching angle 30°; (c) 60°; (d) alternation between 60° and 15°; (e) three-dimensional view of the consequences of hyphal branching. The arrow indicates when the direction of growth has become 180° or more to the direction of growth of the hypha emerging from the spore (germ tube).

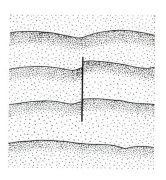

Figure 2.2: The importance of lateral contact between hyphae to synchronize rhythmic behaviour at the mycelial front (production of denser hyphae, represented by denser stippling). Interruption of the contact by a cover slip (black bar) brings a lack of synchrony, which is recovered when the the front passes the cover slip. Redrawn from an original by H. Nguyen Van.

rhythmically growing colony, the synchronization between the hyphal branching system is interrupted by preventing interhyphal contact through anastomoses.

In the (ideal) colony described above, the only differentiation that occurs is the result of the increasing age of the hyphae as one moves to the centre of the sphere. However in the natural world, in a heterogeneous environment, there will be gradients of hyphal density or differentiation within the colony.

2.3 Spread on and in artificial media

The interplay of hyphal extension and hyphal branching is well demonstrated by colonies spreading on the surface of nutrient-agar or, as described below, in nutrient solutions. Initially, colony growth is a stochastic process. By this, we mean that there is a degree of variation in branching angle and distance between the branches. Furthermore, fungal hyphae have a tendency to grow in a spiral, which is right- or left-hand according to the species (*Figure 2.3*). There is variation of the extent to which hyphal growth diverges from the straight line, depending on the extent of the restraining friction of the medium acting against the rotational force within the hypha (possibly brought about by the nature of the molecular architecture of the wall). The extent to which variation in all the various properties identified above is expressed will depend on the local environment in the agar. Suffice it to say, having values for measured variation in the above properties, it is possible to simulate the early growth of a colony by computer with considerable accuracy (*Figure 2.4*).

As the colony grows across the nutrient agar in its mature circular shape, the hyphae grow into and use the nutrients in the previously uncolonized medium, but for the older parts of the mycelium, the medium becomes increasingly depleted of nutrients. Let us take oxygen supply as an example of an environmental factor. Its availability on the surface, on the basis of other microbial studies, is limited by hyphal density and in the agar there is almost certainly a sharp gradient to the lower layers (*Figure 2.5*),

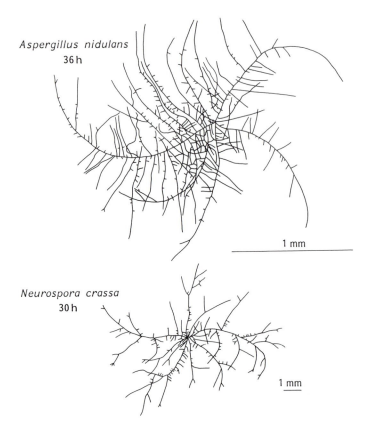

Aspergillus nidulans
36 h

1 mm

Neurospora crassa
30 h

1 mm

Figure 2.3: Spiral growth of the leading hyphae of *Aspergillus nidulans* and *Neurospora crassa* on agar media. The times are those since the culture was started. Reproduced from ref. 4 with permission from The British Mycological Society.

which have been depleted of the gas by the high temperature to which the agar is subjected in preparation, and into which oxygen is diffusing slowly. This oxygen gradient immediately leads to a gradient of decreasing hyphal density from the agar surface, as shown in *Figure 2.6*. Also it is likely that it is oxygen limitation that causes the very regular position of hyphae at the margin of the colony, particularly in one in which there is a significant density of hyphae. In general, each hypha is equidistant from its neighbours. Of course, it could be that a branch is moving along a line represented by the maximum concentration of other nutrients in the agar. However, apart from a small group of lower fungi, there is no evidence that hyphae are chemotropic to organic nutrients (though the situation with respect to inorganic nutrients is not completely clear). A more compelling explanation, based on studies on the orientation of hyphae originating from spores distributed within a liquid medium, is that it is best explained in terms of the oxygen tension within the medium.

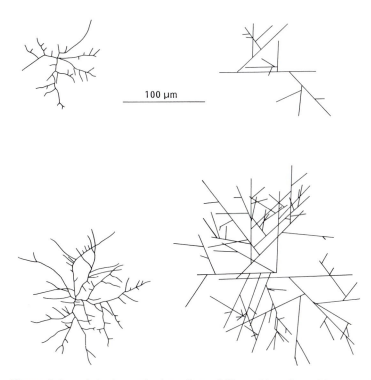

Figure 2.4:   Actual growth of a colony of *Mucor hiemalis* and its computer simulation after 5 h (above) and 7.5 h (below). Tracings of photomicrographs (left) and computer simulation (right). Reproduced from ref. 1 with permission from the British Mycological Society.

There is little doubt that oxygen can become an important limiting factor in three-dimensional growth. We see such growth best in agitated liquid culture. The inoculum for such culture is either spores or fragmented mycelium (used for fungi which do not sporulate readily and produced by subjecting mycelium to a short burst in a blender). After inoculation, if the medium is uniformly agitated at a suitable speed, balls or 'pellets' of mycelium are formed (*Figure 2.7a*). When the pellets reach a certain size, the interior begins to break down (autolyse) and large pellets can be devoid of structure in the more internal parts (*Figure 2.7b*). Autolysis is due to lack of oxygen. The diameter of the pellet at which autolysis begins to take place depends on the density of its hyphae. If there is a high density, the radius will be around 2 mm, the distance being that at which the concentration of oxygen in the centre of the pellet becomes zero, as a consequence of the rate of diffusion of oxygen from the medium being unable to cope with the metabolic requirements at the centre. A radius greater than 2 mm is achieved in those pellets that have less dense hyphae (and a more open mycelium), allowing for convective (turbulent) movement of medium through the pellet.

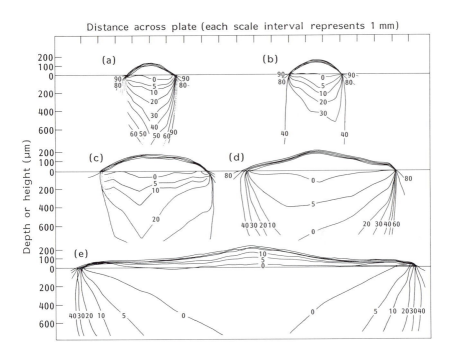

Figure 2.5:   The distribution of oxygen in and around colonies of the bacterium *Escherichia coli* at different ages. (a)18 h; (b) 24 h; (c) 44 h; (d) 68 h; (e) 168 h. Reproduced from ref. 2 with permission from the Society for General Microbiology.

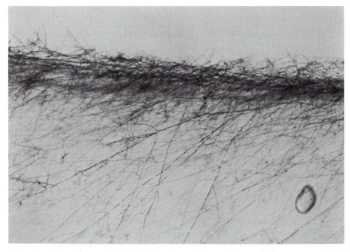

Figure 2.6:   A vertical section through a colony of *N. crassa* growing on nutrient agar showing the decreasing hyphal density in the agar with depth, presumably a consequence of the decreasing oxygen concentration. Photograph by M. Shasavani.

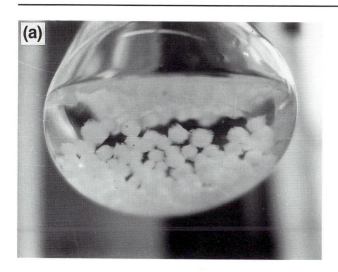

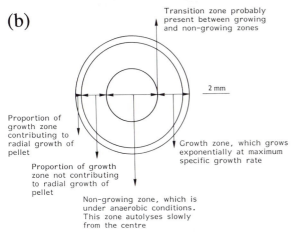

Figure 2.7: (a) Mycelial pellets of *Aspergillus ochreaceus* growing in liquid culture. Photograph by H. Neumeister. (b) Diagram showing the different regions in a growing mycelial pellet. The dimensions of the various regions depend on the fungus and the growth conditions. Reproduced from ref. 3 with permission from Springer-Verlag.

2.4 Across the spatial divide

When a fungus spreads out from a food source on to regions devoid of the nutrient or nutrients required for growth, there must be movement of that nutrient or nutrients within the hypha to allow it to extend at the tip. For extension over very short distances (a few mm), the movement of the nutrient within the aqueous phase of the hyphal cytoplasm can be by diffusion. However, as extension proceeds on the nutritionally deficient surface, diffusion will be inadequate to supply the materials over the distance reached by an extending hypha and at the rate by which the distance has been achieved. Thus movement of nutrients must be by a mechanism other than diffusion. There must be a further input of energy (over that generating any diffusion gradient) to maintain, at the hyphal apex, the requisite nutrients at the concentration values necessary for continuing extension. The process demanding this additional input of energy, by which nutrients are shifted from the site of absorption to another part of the mycelium, is termed 'translocation'.

Translocation within the mycelium is, where studied, predominantly by water flow, though there is some evidence that, in certain circumstances, there can be movement of nutrients in vesicles over significant distances as a result of contractile mechanisms. Water flow is generated by the uptake of nutrients (particularly carbohydrates, from what little evidence we have) by the mycelium on the nutrient substratum such that the hyphae have a lower water potential than the external medium (*Figure 2.8*). In consequence, water flows into the hyphae and the hydrostatic pressure so generated drives a flow of solution towards the mycelial front where growth is occurring. The volume flow is dissipated at the front by the increase in volume of the hyphae (extension growth) and the production of droplets at the hyphal apices (*Figure 2.9*). The droplets have a lower osmotic potential than the hyphae or that of the nutrient substrate from which the mycelium grows, which means that essentially the water leaves the cytoplasm ultrafiltered by the plasmalemma of many of the nutrients in the translocation stream.

The extent to which nutrient movement can be brought about by water flow will depend on the extent to which water lost from the mycelium as the nutrient solution is in transit to the growing front. Water will be lost by evaporation or by absorption by the substratum over which the mycelium has been growing. In either case, the loss of water will lead to the dissipation of the hydrostatic pressure gradient.

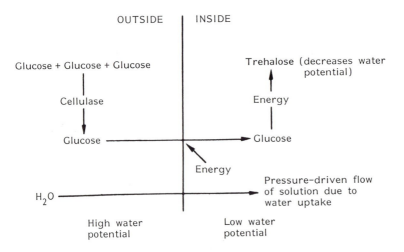

Figure 2.8:   Diagram of those processes that underlie the bulk flow of solution through the mycelium of *Serpula lacrymans* growing from the cellulose in wood over a non-nutrient surface.

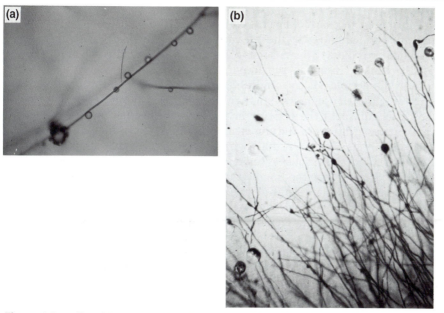

Figure 2.9: Droplets produced on the surface of fungal hyphae: (a) *Botrytis cinerea* growing on nutrient agar. Photograph by K. Blum. (b) *S. lacrymans* growing on Perspex. Photograph by C.R. Coggins.

2.5 Rhizomorphs

In the higher fungi, basidiomycete species particularly, the mycelium becomes differentiated into long linear structures called rhizomorphs which are involved in the spread of these species from one substrate to another. Rhizomorphs can differ significantly in their developmental pattern and in their internal organization. In the mature state, many are characterized by having wide diameter hyphae (vessel hyphae) that form a channel running along the organ, through which translocation can take place, since there is a much lower resistance to solution flow than in more normal hyphae (*Figure 2.10*). These vessel hyphae are found in the inner region of the rhizomorph. The outer region is often composed of closely packed hyphae with thick melanized walls. It is likely that this layer minimizes water loss into the external environment.

Rhizomorphs are produced by fungi living in woodlands and forests and serve as a means of invading pieces of fallen timber from pieces already colonized. Rhizomorphs are also produced by some ectomycorrhizal fungi (see *Figure 3.7d*, p. 41), allowing the colonization of the roots of previously uninfected seedlings. Indeed, the rhizomorphs can act as a conduit for the supply of carbon compounds from a mature tree to a seedling, which is insufficiently able to generate its own photosynthate due to being in conditions of poor illumination, or from another seedling which is located more favourably. Perhaps the best known example of the translocatory ability of rhizomorphs are those produced by the dry-rot fungus *S. lacrymans*. The rhizomorphs allow this particular fungus to spread round a building with disastrous effects on the integrity of the timber, in which the fungus causes brown rot. Interestingly, this fungus is very susceptible to dry conditions. Yet the fact that the organism can transport water through its mycelium and this water is eventually lost to the environment, particularly at the growing mycelial front, means that the fungus can itself alter the water relations of the environment to make them more favourable. Of course, that will only be the case if the environment in which the fungus is growing can be considered to have restricted volume in terms of of its ability to absorb water. A large volume will mean that the fungus will be unable to alter the water potential of the environment to any significant degree. Thus for the dry-rot fungus, its ability to grow in the interface of building materials, for instance between plaster and stone or behind panels, means that the fungus finds little difficulty in providing for itself an appropriate environment in terms of water relations (*Figure 2.11*).

Finally, it must not be assumed that, when a rhizomorph traverses a woodland under the leaf litter layer from one piece of fallen timber to another, the rhizomorph will not make use of the nutrients available in the litter and associated soil. Of course, such

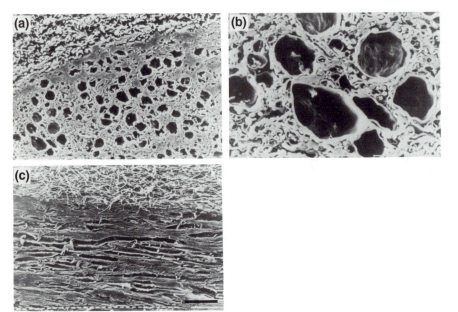

Figure 2.10: Scanning electron micrographs of cross- (above and centre) and longitudinal (below) sections of rhizomorphs of *Armillaria mellea* sensu lato. The hyphae with by far the widest diameters (vessel hyphae) in the central region (medulla) of the rhizomorph are presumed from indirect estimates to be the channels for translocation. Note in the longitudinal section how the vessel hyphae run uninterruptedly along the rhizomorph. The bar represents 100 μm; (c) and (b) respectively at 2x and 7x higher magnification. Photographs by J.W.G. Cairney.

an assumption might seem warranted in view of the presence in the rhizomorph of the impermeable outer layer described in the previous section. However, a rhizomorph is capable, in the presence of an ample supply of water, of putting out hyphae into the surrounding environment (*Figure 2.12*). Although there is no experimental evidence to support the idea, it is hard to believe that such hyphae do not act in an absorptive capacity.

2.6 Translocation in other types of mycelia

Virtually all the studies on long-distance translocation of nutrients in fungi, which have provided the evidence for the mechanism being one of mass flow of solution, have been carried out on three fungi, namely *S. lacrymans, Morchella esculenta* (in relation to sclerotial formation) and, to a lesser extent, on *N. crassa*. However, there is little doubt that pressure-driven flow of solution must occur in a wide range of fungi. This is because the presence of droplets on mycelia is ubiquitous. One can see them on the aerial parts of mycelia on agar and they are seen often on rhizomorphs of species other than *S. lacrymans*. Droplets are often characteristic of the caps of mushrooms and toadstools.

If a pressure gradient can be used to drive the flow of water through mycelium, it must be apparent that, provided the water can be replaced at the necessary rate,

Figure 2.11:   Mycelium of the brown rot fungus *S. lacrymans* growing in a building, from wood (bottom left) through and on stone. Photograph by C. R. Coggins.

evaporation should be able to generate an equivalent movement. This has been shown to be the case for a mushroom, in this case the fruiting body of *Polyporus brumalis*. On the other hand, it must be realized that, if there is an unfavourable water potential at the source of the nutrients, that is, the potential is much lower outside than inside, there will be a reversal of water flow in the mycelium. Growth at the mycelial front will continue only as long as there are nutrients within the protoplasm in the vicinity of the hyphal apex. The water for growth can be drawn upon from the droplets residing on the wall in that region of the hypha. In a short time, however, such a reservoir of water will be exhausted and growth will cease. Speculating a little, one can see that water flow in different parts of a mycelium is controlled by osmotic gradients and any resistances to that flow could be an important determinant of mycelial development. Thus, in the larger fungi, at least, water availability may be significant in helping the fungus develop its specific morphology.

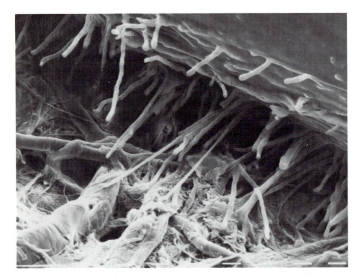

Figure 2.12: Scanning electron micrograph of hyphae extending from a rhizomorph (upper right) of *A. mellea* sensu lato on to the surrounding substrate (moist filter paper, lower left). Bar represents 10 µm. Photograph by J.W.G. Cairney.

References

1. Hutchinson, S.A., Sharma, P. and Clarke, K.R. (1980) *Trans. Br. Mycol. Soc.*, **75**, 177–191.
2. Peters, A.C., Wimpenny, J.W.T. and Coombs, J.P. (1987) *J. Gen. Microbiol.*, **133**, 1263–1275.
3. Trinci, A.P.J. (1970) *Arch. Mikrobiol.*, **73**, 353–367.
4. Trinci, A.P.J., Saunders, P.T., Goswami, R. and Campbell, K.S. (1979) *Trans. Br. Mycol. Soc.*, **73**, 283–292.

Chapter 3
The substrate

3.1 Introduction

Cultures on agar and in agitated liquid systems are entirely artificial systems, which, while they are excellent models and are indispensable for research and other purposes, do not necessarily resemble fungal mycelia in their natural environments. In the latter, mycelial growth is always dictated by the immediate surroundings, the habitat, which provides space and nutrients (including water and oxygen) for the development of the mycelium. This habitat, or substrate, as it is called in mycology, is often highly specific for a given fungus. Furthermore, the interaction between fungus and substrate is such that the two together must be regarded as integral parts of a unified system. In this chapter, we consider this interaction, concentrating mainly on phytopathogenic fungi, though other examples are also referred to.

3.2 The substrate

The substrate can be defined as the space in which a fungus grows as well as the material that fills the space and provides the nutrients for growth. This definition does not include any indication of size, which can be as small as a single cell, for example, of an amoeba parasitized by a fungus, or very large, as the trunk of a large tree or a wide area of soil in a wood or in a lowland meadow. Equally, the definition takes no account of the materials that make up the habitat.

The habitat is fundamentally heterogeneous in three respects:

(i) It can consist of *different materials* close together, that is, it is patchy and differing widely in nutrient content and density which might affect the movement of gases, including water vapour. In a trunk of wood, for instance, there are different tissues, consisting of different cells, which themselves differ in amount and type of their constituents. Inside a soil, we find a mixture of small inorganic particles, plant debris of different size and state of degradation, remains of animals, and so on. These small areas are sometimes called microhabitats and normally contain different fungi, each of which has to develop strategies to migrate from one suitable substrate to another through the mainly hostile space and material in between.

(ii)   There is a *multiplicity of gradients* superimposed on the heterogeneity. We have seen how a gradient of oxygen develops in the artificial system of a petri dish containing a fungal culture on agar. A similar situation can develop within a natural substrate. From the surface to the interior of the substrate there will be a gradient of oxygen tension, the steepness of which will be governed by the consumption of the gas by microorganisms; in most cases the oxygen is replaced by carbon dioxide (*Figure 3.1*). Other gradients are formed by light or by temperature, which inside wood or soil differs from the ambient air, which will have the more extreme values, depending on the season.

(iii) There are also *temporal changes* in the substrate. To take the simplest example, light and temperature oscillate daily and during the year. But the fungi themselves will bring about change as the material on which they are living is degraded and consumed as the mycelia grow. On the other hand, new substrate may arrive, like leaf litter on soil, or an increasing number of aphids yielding more exudate. For those fungi living on animals, there may be new hosts as a result of numbers coming together to form a bigger assemblage. Plant-parasitic fungi have to adapt their physiology to any alterations in the physiology of the host during its development during the vegetative phase.

A fungus in its environment is exposed to all these factors and thus has to develop the means and strategies both to use those factors that are advantageous and to cope with those that are disadvantageous. As an example of what may be involved, we consider phytopathogenic fungi. They provide a comprehensible set of illustrations of the issues under consideration. There is also another reason for considering this particular group of fungi, namely the particular problems of invading and colonizing a living substrate. Phytopathogenic fungi, because, for the most part, they are associated with other living organisms, are called 'biotrophic'. There are some which quickly kill their hosts; these are called 'necrotrophic'.

3.3 Plant-biotrophic fungi and their hosts

3.3.1 Introduction

For any fungus there is a race for nutrients; either it is faster or more powerful than its competitors, or it dies. Thus, there is a permanent pressure to collect the combined carbon from the substrate, whether it be dead or living. Indeed, if a living green plant is invaded, and the plant remains alive, there will be a continuing supply of combined carbon to the fungus. In this instance, the fungus has become parasitic of the green plant but in nutritional terms it is a biotroph. In order to be a successful biotroph (and beat the competition), the fungus must have a successful means of finding its host and

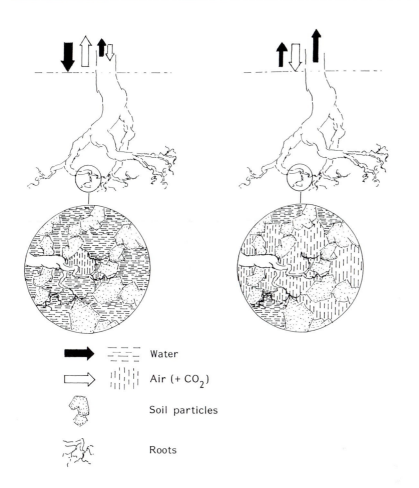

Figure 3.1:   Diagram showing how the ventilation of soil can be altered by the changing water content; altered by rain, evaporation and transpiration and also by movement of air through the root.

of effectively gaining its nutrients once contact has been made. The path is beset with problems.

3.3.2 Finding the host

In the simplest case, a phytopathogenic fungus could arrive on its host carried by wind. This method is very ineffective. The bulk of spores so carried sink to the ground close to their source, that is, an already infected plant. Hosts, if they are plants of the same species, as is usually the case, are normally some distance apart. If these distances are to be bridged with any degree of success, then spore production must be very high. This is the case for rust fungi, which produce an enormous number of infective spores (aeciospores, uredospores and basidiospores) in three out of the five successive generations (see Life Cycle 4 in Appendix A).

A more effective, and better targeted, means of transport is provided by animals, especially insects and birds. They fly from one food source to a similar one carrying the spores with them. Honey-sucking or pollinating insects carry the fungi living in or on a flower, especially the yeasts living in the nectar. Wasps feeding on rotten apples, pears or plums transport the spores of *Monilinia fructigena* (teleomorph: *Sclerotinia fructigena*) to the next fruit and even provide the wound by which the fungus can enter the fruit. The ergot fungus, *Claviceps purpurea*, produces a honeydew containing conidia. Insects attracted by this sweet fluid take up the spores while feeding and transport them to another host plant.

It is even more effective to avoid hunting for a new host but wait until it comes within reach. Smut fungi live for years as soil saprotrophs but when the roots of a host come into the vicinity the fungus will readily invade them. Thus, fields can contain corn smut (*Ustilago maydis*) for up to 12 years in the absence of the host plant. If maize is grown in such a field during this time, it immediately becomes infected.

Fungi attacking animals often use the same principle; they wait at favourable sites until a suitable host arrives. Fungi parasitizing small animals often form spores that can adhere to the cuticle of these animals. This has been shown to be so with nematode parasitizing fungi. The spores of *Drechmeria coniospora* form buds covered

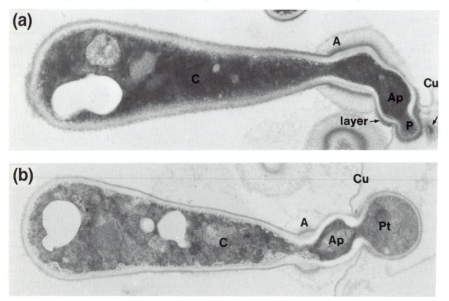

Figure 3.2: Conidia of *Drechmeria coniospora* adhering to a nematode. A, adhesive layer; Ap, appressorium; P, penetration tube; Cu, cuticle of the nematode; layer, possible extra adhesive material. Photograph by J. Dijksterhuis.

with a glue, that fixes the spore to the sensory organs (*Figure 3.2*). Contact for only a very short time is sufficient for the spore to become fixed. In the genus *Nematoctonus* a sticky knob is formed by the germination tube immediately after the conidium has been liberated. The germinated conidium can remain in this state for several weeks until a suitable host comes by. There is no doubt about the effectiveness of these methods, since these fungi produce far less spores than those species that work on the shotgun principle; those that produce non-specialized spores in great numbers so that there will be a chance that some will make contact with their future hosts.

Large animals, including man, are subjected to fungal attack from their immediate close surroundings. It is common knowledge that naked skin can make contact with fungi on submerged structures (especially wooden ones) in swimming pools and saunas. Equally, fungi attack birds from the material in their nests, and moles, foxes and badgers from that lining their holes.

Infections by direct contact between hosts are also common. *Armillaria mellea* is a much-feared plant pathogen because its rhizomorphs are able to grow through the soil from one tree to another; because of this growth through the soil, and since also direct root contact can cause infection, it is very difficult to stop an *Armillaria* infection spreading through a wood or an orchard. In animals, including human beings, dermatophytes spread by direct contact.

3.3.3 Overcoming barriers

When the new host has been reached, the fungus has to reach the nutrients within the living cell. We know plants and animals can defend themselves against infection, because infective propagules or cells can be present on the organism without infection. The defence is organized in a hierarchical manner, as shown in the case of a higher plant in *Figure 3.3*.

The first line of defence can be considered the equivalent to the creation of a glacis before a fort, that is, a means of preventing the fungus from making contact with living cells. The felt-like surface of hairs on a leaf or stem serve this function. Very smooth surfaces can serve a similar function if they cannot be penetrated readily, such that rain or wind is given an opportunity to flush off any spores that have landed on the surface before they have made any progress with penetration.

The next line of defence is to prevent the invader becoming active. Effectively, this means prevention of spore germination. In this respect there are advantages to a smooth, hydrophobic leaf or stem surface, since water, the necessary requisite for spore germination, only resides for a very short time. Equally, those other aspects of leaf morphology that aid dispersal of surface water will be an important element in the defence against fungal attack. Into this category come spines, denticles or prolonged tips, which act rather like gargoyles in dispersing water.

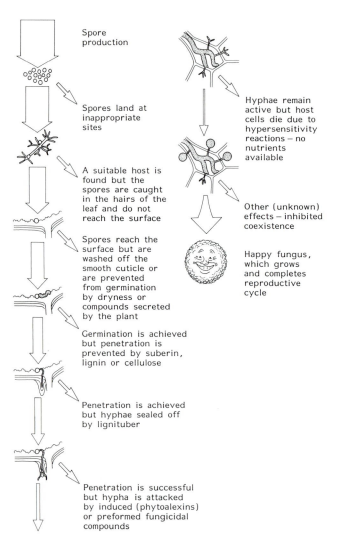

Figure 3.3: Diagram showing the means and/or the organization of the defence of plants against fungal attack. An arrow pointing diagonally represents the successful defence against the fungus; vertical arrows represent the successful overcoming of a defensive barrier by the fungus. Redrawn from an original by E. Schwarzbach.

Even when these barriers have been overcome, there can be further problems facing the fungal spore. Even though it has been able to reside on the leaf or stem surface for sufficient time and with sufficient water to germinate, germination may be inhibited by compounds produced by the plant. The characteristic smell of some leaves is due to substances produced by the glands on the surface. These substances can inhibit fungal spore germination, as has been shown to be so with the blackcurrant, *Ribes nigra*.

Even successful germination does not necessarily spell success. The developing germ tube has to find (indeed maybe force) its way into the plant and into its cells to reach the required nutrients. A common route is via wounds. Larger plants (and animals) always seem to have some small lesions and using them is a simple and safe way to make an entry. The problem for the fungus is to find such a lesion. Fungi using this route of attack are either capable of delaying germination until such a lesion is formed in the near vicinity, or are capable of growing for some time either without nutrients or saprophytically on those nutrients found on leaf surfaces, as is the case with *Alterneria tenuis* or *Botrytis cinerea*.

Other fungi are less opportunistic, having a more clearly defined strategy for entering the plant. Thus the zoospores of *Peronospora* and other oomycetes swim to the stomata in the surface water on leaves formed by dew in the early morning or by rain, and invade the mesophyl via the stomatal pore. The mildews form appressoria as hyphal branches, which at first fix the fungal mycelium to the plant surface then produce enzymes that break through the cuticle and underlying cell wall. More recently it has been shown that the appressoria of *Magnaporthe grisea* fix themselves tightly to the leaf surface and produce tip-growing infection pegs that puncture the cuticle and underlying wall. The process is purely physical, since appressoria can penetrate plastic. The germ tube generates very high turgor [80 MPa (= 80 bars)]. In finding their way to a suitable site for penetration, germ tubes can be guided by the microscopic undulations of the cuticular surface.

As in other instances, plants are able to defend themselves against attacks of the kind just described. The wall can be quickly thickened by the provision of additional material. When this occurs, there is a race between wall penetration by the fungus and synthesis of new wall material, resulting in a cone-like structure, the lignituber (see *Figure 3.3*). Eventual invasion of the cell, if it occurs, activates the next line of defence. There may be production of antifungal substances, which kill the fungus directly. On the other hand, the plant cells in the locality of the attack may die at the very first contact (hypersensitivity reaction), which means that the invading fungus only finds dead autolysed cells which probably do not provide much in the way of appropriate nutrients to make up for the declining nutrients of the invading spore and its germ tube.

Modern research has shown that virtually all plants have evolved some form of defence against fungal attack, but obviously there are plants (and animals) which become infected by fungi, often systemically. All barriers to attack have been overcome. In nature, this is usually the exception rather than the rule. This is because of the permanent race between the continuing evolution of defensive barriers by potential hosts and the evolution by the fungi of new methods of overcoming these barriers. In consequence, there has been the development of specificity of parasitic fungi for their host plants. Thus there is a balance; not all plants are infected and not all fungi are destroyed before establishing themselves within the plant so that they can complete their life cycle.

The details we have described for phytopathogenic fungi demonstrate how fungi can be adapted to their substrate. Turning to saprotrophic fungi in the soil, it should be clear from the above that these fungi will face problems in invading dead plant material, which, if not as difficult to overcome as those facing phytopathogenic fungi which attack living plants, are of a similar kind. Thus the saprotroph will still have to overcome fungitoxic compounds and physical barriers in the plant remains that it is trying to attack, and, of course, there will be competition from other microorganisms, including other fungi. As with the phytopathogenic fungi, there will be a degree of specificity amongst the saprotrophs for the substrates that they can attack.

3.3.4 Fungi as sinks for nutrients

When a fungus manages to become established in the living plant cell it gains ready access to the nutrients therein. To gain this access, hyphal branches (haustoria; *Figure 3.4*) are produced within the plant cell wall, which protrude into the plant cell protoplast (without rupturing it), and by which the fungus absorbs solutes that travel across both the plasmalemma of the plant cell and that of the haustorium. In this way, the fungus has readily consumable nutrients, that is, low molecular weight metabolites, at its disposal and has no need to consume energy in trophic growth or in the production of extrahyphal enzymes. The still living plant replaces this loss of metabolites from non-infected cells, resulting in what is often a severe loss of yield. The fungus has become a sink for nutrients in the plant that it has invaded.

A fungus often alters the physiology of the host. Rust fungi alter the morphology of their hosts. *Uromyces fabae* lives (and produces uredospores) on *Euphorbia cyparissias*. Healthy plants have small, non-succulent leaves and many branches and flowers, whereas infected plants are unbranched monopodia, with light green succulent leaves and do not form flowers (*Figure 3.5*). It is believed that these alterations are caused by the secretion of compounds by the fungus that act as growth regulators. It is also believed that the movement of nutrients within an infected plant towards a haustorium may be aided by the synthesis of compounds by the fungus which cannot be metabolized by the plant.

Frequently compounds produced by the fungus make the plant poisonous to predators or consumers (*Table 3.1*). The best studied case is *Lolium temulentum,* which contains a toxin produced by an endophyte. The toxin affects the central nervous system of grazing animals, the name *temulentum* = trembling or causing trembling, highlights this effect.

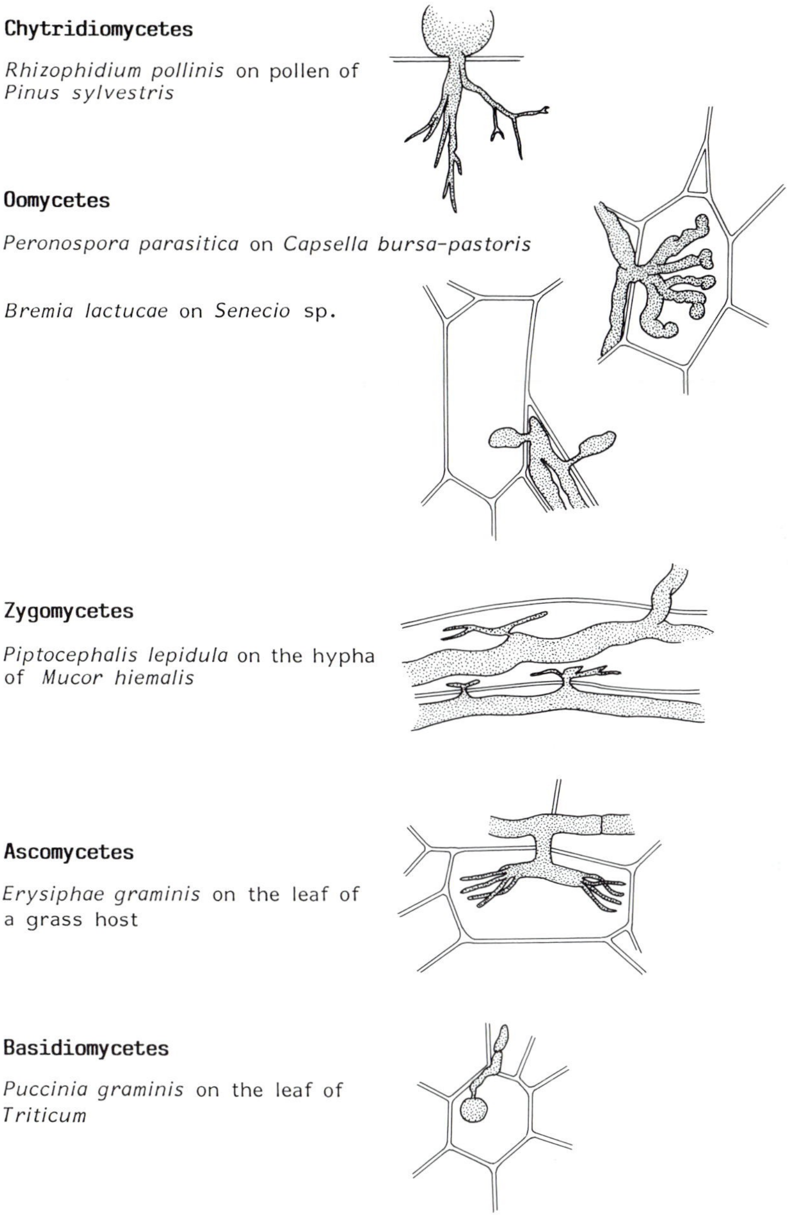

Chytridiomycetes

Rhizophidium pollinis on pollen of
Pinus sylvestris

Oomycetes

Peronospora parasitica on *Capsella bursa-pastoris*

Bremia lactucae on *Senecio* sp.

Zygomycetes

Piptocephalis lepidula on the hypha
of *Mucor hiemalis*

Ascomycetes

Erysiphae graminis on the leaf of
a grass host

Basidiomycetes

Puccinia graminis on the leaf of
Triticum

Figure 3.4:   Forms of haustoria in the various fungal classes.

Table 3.1: Host grasses or sedges reported to exhibit toxicity or increased resistance to grazing when containing hyphae of certain fungi. The term endophyte refers to a fungus the hyphae of which are not associated with any reproductive structure while associated with the plant host. Data by kind permission of K. Clay

Host species	Fungus
Andropogon spp.	Balansia spp.
Cenchrus echinatus	Balansia obtecta
Cyperus virens	Balansia cyperi
Cyperus pseudovegetus	Balansia cyperi
Cyperus rotundus	Balansia cyperi
Dactylis glomerata	Epichloë typhina
Danthonia spicata	Atkinsonella hypoxylon
Festuca arundinacea	Acremonium coenophialum
Festuca longifolia	Endophyte
Festuca obtusa	Endophyte
Festuca rubra	Epichloë typhina
Festuca versuta	Endophyte
Glyceria striata	Epichloë typhina
Lolium perenne	Acremonium lolii
Lolium tementulum	Endophyte
Melica decumbens	Endophyte
Panicum agrostoides	Balansia henningsiana
Paspalum dilatatum	Myriogenospora atramentosa
Paspalum notatum	Myriogenospora atramentosa
Stipa leucotricha	Atkinsonella hypoxylon
Stipa robusta	Endophyte
Stipa mongol	Endophyte
Tridens flavus	Balansia epichloë

This example is at what might be called the edge of pathogenicity; the fungus does not cause any visible harm, although it must depend for its nutrient supply on its host. On the other hand, the fungus gives protection to the host, by reducing the grazing pressure. The loss of resources caused by the fungus could well be lower than the gain brought about by the protection. Here one can see mutual benefit to fungus and plant.

Mutual benefit may be temporal in nature. Strawberry plants, like other rosaceous plants, contain protoanthocyanidines, which can act as antifungal agents. The immature fruit also contain high concentrations of this type of substance and are hence well protected against fungi, especially *B. cinerea,* the grey mould. When the fruit become mature, the protoanthocyanidines are transformed to anthocyanidines, responsible for the familiar red colour of ripe strawberries. Usually the kernels that contain the seeds are dispersed by birds and small mammals eating the fruit, attracted by their red colour. If the fruit is not eaten, it can be attacked by the grey mould which breaks down the flesh, allowing both the release of the seeds and providing nutrients for the germination of the kernels or the seeds.

The most impressive examples of mutual benefit are lichens and mycorrhizas in which both partners gain mutual benefit. Lichens are highly successful 'compound organisms' produced by the association between a few species of unicellular algae or

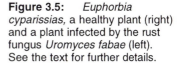

Figure 3.5: *Euphorbia cyparissias,* a healthy plant (right) and a plant infected by the rust fungus *Uromyces fabae* (left). See the text for further details.

cyanobacteria (or in a few instances both) with many species of fungi. Lichens grow throughout the world but are particularly successful in extreme environments. At first sight, it would seem that the fungus gains more from the association than its photosynthesizing partner (through the consumption of the products of photosynthesis, see p. 52; *Figure 3.6*). Indeed, if isolated, the fungus is difficult to grow in culture. On the other hand, there is little problem in having the alga grow independently of the fungus. Nevertheless there is evidence that, when residing in the lichen, the alga becomes much more tolerant of extreme conditions, through its association with the fungus.

Mycorrhizas are the association of fungi with the roots of plants *(Figure 3.7)*. The role of these associations in benefiting the higher plant through an increased supply of phosphorus is described on p. 58. The fungus benefits from the supply of carbohydrate from the host. As much as 30–35% of the photosynthate gained by a beech forest is metabolized by the mycorrhizal fungi.

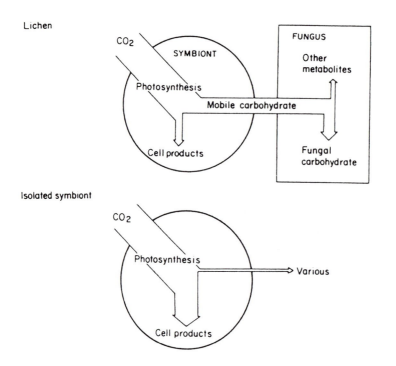

Figure 3.6:   Diagram showing the effect of the removal of the alga *Coccomyxa* from symbiosis in the lichen *Peltigera aphthosa*. After separation from the fungus, carbohydrate release from the alga declines sharply. The width of the arrows indicates the relative amount of carbon moving along each pathway. The experiment demonstrates the ability of the fungus to tap the photosynthetic products within the lichen association. Reproduced from ref.1 with permission from Edward Arnold Ltd.

On the other hand, there are mycorrhizal higher plants that appear to be the sole beneficiaries of such an association. This is so for the orchid *Neottia nidus-avis* which is colourless, not containing chlorophyll *(Figure 3.8)*. Here the fungus exploits the combined carbon in the soil by breaking it down to soluble form, some of which must be used for growth by the orchid. This ability of fungi forming mycorrhizal associations to break down insoluble carbon compounds in the soil is widespread (see p. 50) and the soluble carbon compounds may supplement those coming from the higher plant. In the *Ericales,* there are members (in the Monotropaceae) with mycorrhizal roots somewhat different in morphology from that described by other members *(Figure 3.7c)*. For one such colourless species, *Monotropa,* it has been shown that the fungus forming the mycorrhizal association obtains the combined carbon not from the soil but from a tree with which it forms an ectomycorrhizal association.

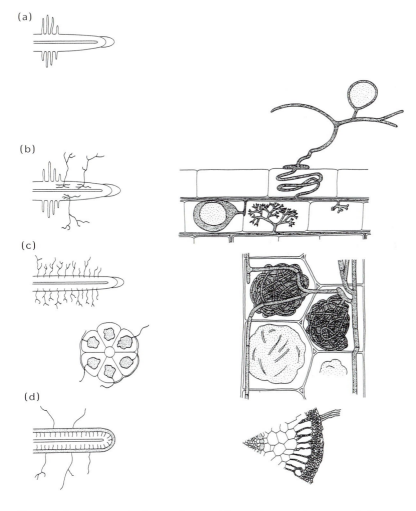

Figure 3.7:   The major forms of mycorrhizas. Diagram of root morphology on the left; the relationship of the mycorrhizal fungus with the individual cells of the root on the right.
(a) Uninfected root with root cap, root hairs and vascular system (stele); (b) vesicular–arbuscular (V-A) mycorrhiza showing, on the right, a spore producing invading hypha and a vesicle (left) and arbuscule (right) inside the cells of the root; (c) ericoid mycorrhiza showing infection hyphae invading the protoplast to form an infection coil and two coils at a later stage undergoing lysis. Also shown is a cross-section of the root, which is hair-like to the naked eye, showing the size of the single layer of cortical cells relative to the stele; (d) ectomycorrhiza of the root of a forest tree, showing the outer fungal sheath and the hyphae penetrating between the root cells to form what is known as the Hartig net. From the sheath delicate rhizomorphs penetrate the soil. The figure is, in part, a redrawing of an original by H. Hudson.

Figure 3.8: *Neottia nidus-avis,* an orchid depending entirely on the nutrients supplied by symbiotic fungi.

Reference

1.   Smith, D.C and Douglas, A.E. (1987) *The Biology of Symbiosis.* Edward Arnold, London.

Chapter 4

Nutrients

4.1 Carbon nutrition

4.1.1 Introduction

All fungi depend on organic carbon; it is the qualitatively and quantitatively most important nutritional element. The bulk of organic carbon on the earth owes its origin to photosynthesis. That organic carbon is a major part of living plants, animals and microbes. Much of their remains are readily broken down, particularly proteins and nucleic acids, indeed these molecules can be degraded by enzymes found in the cells of all living organisms. Lipids are a little more refractory. But it is when the carbon is in the form of structural material, such as cellulose, lignin, chitin and keratin, that the molecules become more difficult to degrade. Fungi, however, are able to degrade all these molecules with great success.

4.1.2 Degradation of plant cell walls

Of the structural material referred to above, cellulose and related polymers and lignin present the greatest challenge as the result of the evolution of massive plant structures such as trees. Since green plants have evolved only a restricted range of enzymes for attacking cell wall material, and then in restricted circumstances (i.e. the abscission layer in the petioles of leaves or fruits), the trunk of a tree presents problems in terms of breakdown because all the enzymes required must be produced by the attacking microorganisms, which must also penetrate the trunk. Chitin and keratin of animals present less of a problem because of their essentially surface location, as does the lignin in the vessels of leaves and young shoots, which is also relatively accessible.

When it comes to the invasion of bulky substrates, there is little doubt that a fungal hypha is better able than a single-celled organism to penetrate into the interior of the substrate. It is not appropriate to go into reasons here, the reader will become aware of them at later points in the text.

The ability to attack the cellulose of green plant walls is widespread amongst fungi, although the ability of an enzyme to degrade the major component, cellulose, is limited by the crystalline nature of the polysaccharide *(Figure 4.1)*. Nevertheless fungi have evolved efficient systems for degrading cellulose. In spite of the large number of cellulolytic fungi and the considerable research on the enzymes that they produce, there is still some uncertainty as to how cellulose is degraded biochemically. The picture emerging, particularly from studies on *Trichoderma reesei,* which is used industrially to degrade cellulose, is that there seem to be two possibilities for the process of degradation:

(i)   The initial attack is by endoglucanases, which break the cellulose at the amorphous regions, followed by the combined action of cellobiohydrolases (which break off cellobiose units from the end of the cellulose chain) and endoglucanases, with the final hydrolysis of the small oligosaccharides to glucose being mediated by cellobiase.

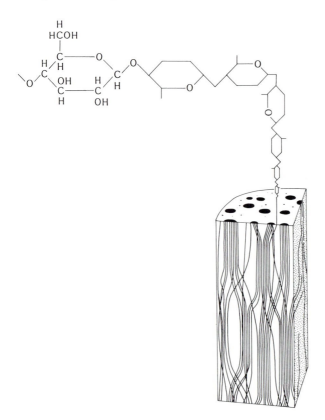

Figure 4.1:   Diagram showing the relationship between the glucan (glucose before polymerization) units forming a cellulose chain and how that chain is associated with other cellulose chains within a microfibril. Where the chains run parallel to each other (solid circles in cross-section), the region is crystalline; where the chains diverge, the region is said to be amorphous.

(ii)   The combined action of cellobiohydrolases is sufficient to yield complete hydrolysis of cellulose. In this case the endoglucanases act only on the solubilized material, such as cellodextrins, to yield cellobiose. This cellobiose, together with that produced by the action of cellobiohydrolases, is converted to glucose by cellobiase.

The preservation of the remains of living organisms in peat testifies to the major requirements of organic decay for oxygen and nutrients. Thus a log on the forest floor can only be attacked from the surface inwards, unless there is some way of allowing more ready access of oxygen to the interior. In woody tissues, the deposition of lignin in the cell wall makes it refractory to breakdown. Essentially, any microbe attacking wood has to gain access to the substrate which will eventually provide the source of energy for growth, namely the cellulose and hemicelluloses, by first breaking down the plastic (polymeric) matrix of lignin in which the polysaccharide is embedded.

Lignin is a generic name for the complex polymers built up from phenyl propenoid units *(Figure 4.2)*. Unlike most other polymers found in nature, lignin is not a well-defined molecule because there are not systematically repeated units. The great majority of bonds are covalent and of considerable variety and are equally in all three dimensions. Furthermore, the molecule is highly hydrophobic. Thus systems required

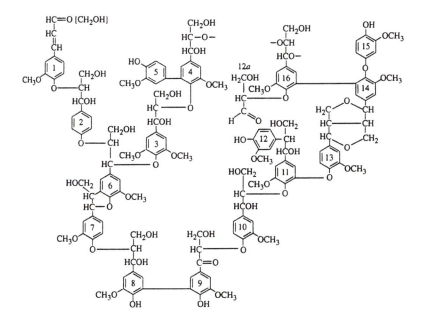

Figure 4.2:   Schematic formula for a portion of spruce lignin consisting of 16 different units. Reproduced from ref. 1.

to break down the molecule must be non-specific and, for the most part, non-hydrolytic as well as extracellular. The most rapid and extensive degradation of lignin is brought about by fungi. Basidiomycete fungi are the most effective but some ascomycetes and fungi imperfecti can cause significant decay. While bacteria can break down the polymer, they are nowhere near as effective.

Lignin breakdown unlocks the polysaccharides within the wall for degradation and utilization by microorganisms, but we should not forget the amount of carbon deposited in lignin itself. It has been estimated that the annual production of terrestrial biomass on the earth is 10×10^{10} tonnes. Of this, approximately 20×10^9 tonnes are lignin. So, while fungi do not appear to make much use of lignin as an energy or food source, breakdown of the polymer is ecologically highly significant. Equally, an understanding of the process is important biotechnologically, since the unlocking of carbohydrate polymers from lignin allows them to be utilized either directly, in the manufacture of paper, or indirectly, through conversion to ethanol. It is the biotechnological aspects that have underpinned recent research on lignin degradation. Particular attention has focused on lignin breakdown by the basidiomycete *Phanerochaete chrysosporium,* especially a strain capable of growing at elevated temperatures.

Lignin breakdown is an aerobic process. It has not been shown to occur anaerobically, which is not surprising, since the process must be oxidative. Further, from what has been said above and because when fungi break down lignin there are a variety of products, any enzymic process must be non-specific and able to bring about the diversity of degradative reactions, which are known to take place. The key enzymes are two types of peroxidase. Considering the one first discovered, lignin peroxidase (ligninase), the enzyme breaks oxidatively carbon–carbon bonds between aromatic nuclei and catalyses a single-electron oxidation of the various aromatic moieties of the lignin molecule so produced *(Figure 4.3)*. The chemistry of ligninase action is not simple. The enzyme cleaves bonds oxidatively and also produces radical cations, which can themselves act as oxidants. Also there is a natural secondary metabolite, veratryl alcohol, which is produced independently of lignin breakdown by *P. chrysosporium* and which ligninase can oxidize to a radical cation which itself is capable of oxidizing lignin *(Figure 4.4)*. This means that it will not matter if the enzyme, because of its size, cannot penetrate the wall; the radical cation of veratryl alcohol or other small molecular weight compounds can act as small diffusible mediators of ligninase activity. The other peroxidase (manganese-activated) acts in a similar manner to ligninase. The hydrogen peroxide required for lignin breakdown could be supplied by a number of enzymes, mostly extracellular, which have been demonstrated to be produced under the requisite conditions by cultures of *P. chrysosporium*. Finally, the fungus needs energy when it breaks down lignin; it seems unable to use lignin by itself to produce new biomass, that energy comes from the polysaccharide within the plant cell wall. The mechanism of lignin breakdown by other fungi seems to be similar to that which has been described here.

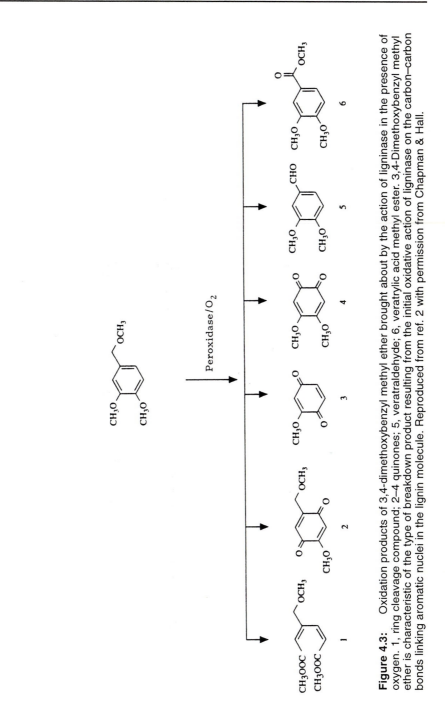

Figure 4.3: Oxidation products of 3,4-dimethoxybenzyl methyl ether brought about by the action of ligninase in the presence of oxygen. 1, ring cleavage compound; 2–4 quinones; 5, veratraldehyde; 6, veratrylic acid methyl ester. 3,4-Dimethoxybenzyl methyl ether is characteristic of the type of breakdown product resulting from the initial oxidative action of ligninase on the carbon–carbon bonds linking aromatic nuclei in the lignin molecule. Reproduced from ref. 2 with permission from Chapman & Hall.

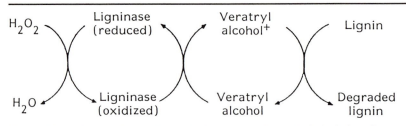

Figure 4.4:   The oxidation of lignin by ligninase with veratryl alcohol as the mediator. Veratryl alcohol radical cations are generated by ligninase and the cations mediate the oxidation of lignin by functioning as one-electron oxidants.

Complete degradation of lignin is not essential for fungi to be able to attack the cellulose in the wall. A number of basidiomycetes only partially decompose lignin. Under these circumstances the bulk of the lignin remains after fungal activity has ceased. Since the remaining lignin, which is oxidized, is brown, the fungi causing this kind of wood decay are called brown rots. Those basidiomycetes that break lignin down completely, because there is little change of colour with decay, are called white rots. Brown rot and white rot fungi are often closely related taxonomically.

Thus far, we have been describing in molecular terms how fungi attack cell walls. In morphological terms the attack takes different forms and is best considered under three main headings *(Figure 4.5)*, as follows.

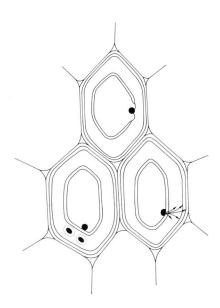

Figure 4.5:   Diagram of wood cells in cross-section, showing the form of attack of fungi causing: white rot (top, showing erosion of wall); brown rot (bottom right, degradation without much obvious change in form); soft rot, cavities produced running parallel to the long axis of the cell. Black circles represent hyphae. Redrawn from original by R.A.P. Montgomery.

Figure 4.6: Photograph of white rot in a stump of *Acer* sp. (fungus unknown), showing the typical fibrous decay of the wood.

(i) *White rot.* There are two distinct classes *(Figure 4.6)*. Either the lignin and polysaccharide are broken down together or the lignin is removed selectively before polysaccharide breakdown. In the first instance the hyphae produce grooves in the wall which is gradually reduced in thickness. In the second case, selective lignin removal results in a variety of appearances at the macroscopic level, but essentially there can be pockets of decay in the wood or it has a mottled appearance. With advancing decay, the wood is still whitish in colour and becomes fibrous in texture. What happens at the microscopic level mostly depends on the fungus and the detailed anatomy of the wood. Further, the two types of lignin breakdown, simultaneous and selective, may occur in the wood at the same time.

(ii) *Brown rot.* With decay, the wood takes on a brown colour, as described above, and becomes friable and cracks cubically, a familiar sight to anyone subject to an attack of dry rot in their house *(Figure 4.7)*. The whole cell wall is degraded, not

Figure 4.7:   Brown rot of building timber caused by the dry rot fungus *Serpula lacrymans*. The mycelium has been peeled away from the wood to reveal intense cracking of the wood brought about by the removal of the water-attracting cellulose and leaving the water-repelling lignin, causing the wood to appear as if it has 'dried out'. Photograph by R.W. Clarke.

just in the vicinity of the hyphae. As indicated earlier, brown rot fungi do not possess a ligninase. The polysaccharide of the wall is broken down and the lignin matrix nearly undigested but not unchanged. Agents that cause decay diffuse into the wood so that the process is not localized in the vicinity of the hyphae. In the early stages of decay, there is a considerable loss of mechanical strength of the wood but little weight loss. It is believed that in these early stages the cellulose is made more accessible to attack by Fe^{2+} (though it is not clear how iron in this state of reduction is produced) together with hydrogen peroxide (Fenton's reagent) to bring about the oxidative degradation of the cellulose which, at a later stage, can then be broken down further enzymically. The Fe^{2+} and the hydrogen peroxide are more readily able to diffuse through the wood than enzymes and the same compounds might also change the structure of lignin by oxidation.

(iii)   *Soft rot*. These fungi break down cellulose and hemicelluloses but do not degrade lignin appreciably. Essentially, soft rots occur in wood exposed in aquatic environments and attacked by fungi of the ascomycetes and deuteromycetes. In longitudinal section the soft-rotting wood is characterized by chains of enlarged cylindrical cavities with (in sections) angular ends (*Figure 4.8*). These cavities are brought about by rhythmic growth of hyphae within the wood.

4.1.3 Utilization of the carbon in protein

While it is clear that plant cell wall material is a major source of carbon for saprotrophic fungi, in organic soils the presence of combined carbon in the form of protein is also significant. Many fungi have been shown to hydrolyse proteins in

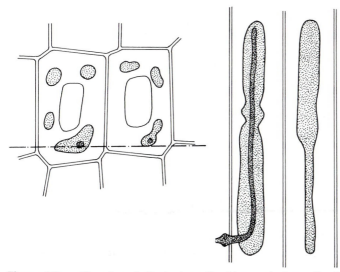

Figure 4.8: The sites of attack of a soft rot fungus in the walls of two wood cells (hyphae densely dotted; cavities lightly dotted). The dashed line shows the line of the longitudinal section.

culture. It is not necessary for there to be complete digestion to amino acids; many fungi can assimilate simple peptides. Of course the breakdown of protein provides an important source of nitrogen; but it can also be an important source of carbon. The use of protein in this way has very considerable significance in the biology of mycorrhiza. The two types of mycorrhiza for which the utilization of protein has been investigated are those of the ericoid type in the Ericaceae, for example, *Calluna, Erica, Rhododendron,* and the ectomycorrhiza of forest trees, for example, *Betula.* The ericaceous types are found in heathland communities which dominate the cold and wet regions of the world (*Figure 4.9*). The roots of the Ericaceae growing in these heaths have very fine, so-called 'hair-roots'. The cortical cells of such roots are filled with mycelium which extends on to the surface to form a loose and weakly developed weft on the surface, with very few hyphae running out into the surrounding soil (see *Figure 3.7c,* p. 41). Amongst their physiological capabilities, these mycorrhiza are able to hydrolyse protein by the secretion of an acid protease.

It has been shown that in culture and in nature many basidiomycete fungi that are ectomycorrhizal with forest trees, that is, which form a sheath of mycelium covering the root and penetrating the cortex (see *Figure 3.7d,* p. 41), are able to utilize peptides and proteins, not only as sources of nitrogen but also as sources of carbon. Experiments have been carried out with mycorrhizal seedlings of *Betula pendula* in which they were supplied with ^{14}C-labelled protein. The carbon from the protein was absorbed and translocated to the leaves. Calculations of the net gain of carbon from the heterotrophic source showed that over the experimental period of 55 days the growing plant received as much as 9% of its carbon from protein. Many non-mycorrhizal fungi, including the common mushroom, *Agaricus bisporus,* can also utilize protein.

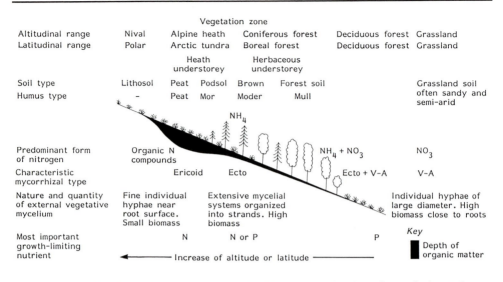

Figure 4.9: The relationship between mycorrhizal types and climatic regions: nival, growing under snow; tundra, northern treeless zone having permanently frozen subsoil; boreal, snowy winters and short summers. V-A, vesicular–arbuscular mycorrhizas; ECTO, ectomycorrhizas. Reproduced from ref. 3 with permission from the British Mycological Society.

4.1.4 Biotrophic carbon utilization

We have seen earlier how fungi exploit living organisms for their food, either killing them first, or relatively soon after they have been attacked, or allowing them to live in order to produce a continuing supply of food. This gradation from early death of the host to its continuing survival is well demonstrated by green plants. In those cases where the fungus is necrotrophic or near-necrotrophic, the fungus obtains its nutrition by utilizing the soluble compounds directly, followed by enzymatic breakdown of the insoluble compounds. If the host continues to live after fungal invasion, it is because the fungus has formed an intimate relation between its metabolism and that of its host, such that there is a continual supply of combined carbon compounds to the fungus from the plant. The most outstanding instances of such a situation, and for which we have clear evidence, are in lichens and ectomycorrhiza. In these associations it has been well demonstrated that photosynthate is continuously moved into the fungus from the green plant. In the fungus, the photosynthate is converted into other compounds, particularly mannitol (and, in the case of mycorrhiza, trehalose as well).

Experiments in which $^{14}CO_2$ has been fed to a thallus of the lichen *Peltigera polydactyla* in the light has shown there is a very rapid movement of photosynthate to the fungus (detectable within 90 sec of the start of the experiment; *Figure 4.10*). The photosynthate that moves depends on the cyanobacterium/alga present. If this is

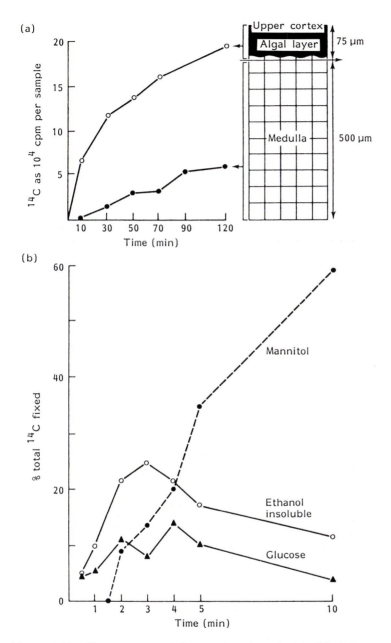

Figure 4.10: The movement of photosynthate from the 'algal' (strictly cyanobacterial) layer to the fungus in the lichen *Peltigera polydactyla*. (a) Movement from the algal layer to the fungal medulla. Discs of lichen were floated in the light on solutions of sodium bicarbonate labelled with ^{14}C. At intervals, the discs were removed and the medulla separated and its ^{14}C content determined. (b) The pattern of incorporation of ^{14}C into the major chemical fractions in the medulla. Reproduced from ref. 4 with permission from Edward Arnold Ltd.

a cyanobacterium, it is glucose which moves. For algae, the compound depends on the genus. Erythritol, ribitol and sorbitol have been shown to be the mobile products of photosynthesis in the lichens in which algae reside. A very significant feature of the release of the mobile photosynthetic product from the photobiont is that it is not demonstrated by the same organisms when they have been isolated from the fungus and are in pure culture (see *Figure 3.6*, p. 40). So a key feature of the association with the fungus is the preferential movement of photosynthate from the phototroph to the fungus rather than to cell products in the phototroph itself.

4.2 Nitrogen nutrition

Fungi are able to use a range of nitrogen compounds (*Figure 4.11*). We have already seen that proteins in the soil can be an important source of both carbon and nitrogen. In general, there is a hierarchy of compounds, such that when a fungus is presented with a mixture, certain nitrogen compounds are used preferentially. To do this, there must be mechanisms for allowing uptake and metabolism of the preferred compound, while preventing the same for the less-preferred compound(s). Also, when the preferred compounds are not available, there must be mechanisms to allow the uptake and metabolism of the less-preferred compounds. To cope with the various eventualities, the following mechanisms have been identified as being likely to be widespread.

(i) There can be competition for transport systems (permeases). This has been well demonstrated for amino acid transport. Within a particular fungus, and this is

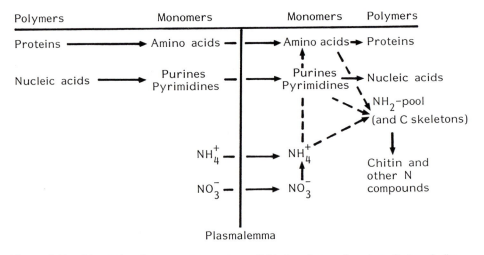

Figure 4.11: The major nitrogen compounds available to a fungus in nature (but excluding chitin) and the general pattern of their metabolism once inside. Dotted lines have been used in some instances for ease of representation of metabolic steps.

certainly so for *Neurospora crassa (Table 4.1),* there are a number of transport systems with different characteristics, for instance whether the amino acids are acidic, basic or neutral. Within the group of amino acids transported by one system, there is a hierarchy according to the affinity of the transport system for the particular amino acid. If one amino acid has a higher affinity than another in the group, it is the one with the higher affinity which will be absorbed preferentially into the hypha from a mixture of the two amino acids. Furthermore, if the one which is preferentially absorbed is at a significant concentration in the protoplasm, uptake of the other amino acid will be inhibited, even in the absence of the internal amino acid in the external medium.

(ii) Ammonium, certainly in those cases where it is the preferred nitrogen source, represses the synthesis of those enzymes associated with the assimilation of amino acids and other nitrogen compounds. It is not ammonia itself that is the effector compound but a metabolic product, probably glutamine, that represses the synthesis of the relevant enzymes. This regulatory system, which operates at the level of gene transcription, has been termed nitrogen catabolite control. Regulation is not only through repression of the synthesis of enzymes for a particular pathway but there is also what is termed inducer exclusion, that is, nitrogen sources other than ammonium are unable to enter the fungus. This means that certain transport systems are repressed. Thus, in *N. crassa,* the general amino acid permease, which can transport 20 natural substrates, including all the L-amino acids which are constituents of proteins, is repressed by the presence of ammonium.

(iii) While, as indicated, there can be repression of both permeases and metabolic pathways for a particular nitrogen compound depending on what other nitrogen compounds are available, equally when that compound is the only source of nitrogen there is induction of the relevant permeases and metabolic pathways for its effective utilization.

Table 4.1:   The major amino transport systems in *Neurospora crassa.* The individual amino acids that are listed are those having the highest affinity for the system. If there is another amino acid in the external medium with similar properties to, and at the same concentration as the one listed, it will be the one listed that will be transported into the hypha

System I: L-neutral amino acids	System II: D- or L-basic, neutral and acidic amino acids[a]	System III: L-basic amino acids	System IV: D- or L-acidic amino acids	System V: L-methionine[b]
Amino acid with highest affinity for the system				
L-Phenylalanine L-Tryptophan	L-Arginine Glycine L-Leucine	L-Arginine L-Lysine	L-Cysteic acid L-Aspartic acid	L-Methionine

[a]   General amino acid transport system.
[b]   Active under sulphur starvation; methionine absorbed as source of sulphur rather than a source of an amino acid.

(iv) It has been shown that starvation of cells of baker's yeast, *Saccharomyces cerevisiae,* and mycelia of *Aspergillus nidulans* and *N. crassa* of one of a number of amino acids leads, in compensation, to increased expression of a number of enzymes in several different amino acid pathways. The cross-pathway character of the response has led to it being termed general amino acid control.

(v) When yeast cells are transferred to a new medium with a different nitrogen source which supports growth, there may be disappearance of those enzymes involved in the metabolism of the previous nitrogen source. There is no particular mechanism for bringing about the disappearance. Synthesis of the particular enzymes is repressed and those enzyme molecules that remain are diluted out by growth.

(vi) In fungi, the ability, indeed necessity, to use a range of nitrogen sources means that both catabolic and anabolic processes may have to take place simultaneously. If ordered growth is to occur there must be sophisticated regulation of metabolic activity, particularly when common intermediates exist. In this respect the metabolism of arginine deserves special mention. The metabolism of this amino acid has been studied in some detail in *N. crassa* and *S. cerevisiae.* For these two fungi, it has been shown that, not only is there the normal type of biochemical regulation, acting at the enzyme level, but there is also compartmentation of certain metabolic pathways and metabolites in mitochondria, vacuoles and nuclei.

These features, allowing metabolic flexibility with respect to the variety of nitrogen compounds available to fungi, are only part of a more general metabolic flexibility which is considered further in Section 4.6 (see p. 63).

4.3 Phosphorus and sulphur nutrition

In artificial media, phosphorus is supplied as orthophosphate, which, in acid media and provided that it is not precipitated out by the calcium present, is readily utilized by a fungus for growth. In nature, fungi can have access to phosphorus also as a result of their ability to break down organic phosphorus compounds in the remains of dead organisms *(Figure 4.12)*. Nevertheless it is well known that of the organic phosphorus compounds in humic, including forest, soils, much (up to 80% or more in the extreme) may be in the form of phytates (inositol phosphates), which are not readily degraded, even by fungi.

We need to digress for a moment to consider the partitioning of phosphorus in the soil. There, inorganic phosphorus is in three main fractions, based on the location of the element: the soil solution, the labile pool and the non-labile fraction *(Figure 4.12)*. That phosphorus in the soil solution is present as dissolved orthophosphate; labile phosphorus is the phosphate adsorbed on to surfaces, which is in equilibrium with the soil solution; non-labile phosphorus is in the insoluble fraction, for the most part in the form of calcium, aluminium and iron phosphate, the calcium phosphates constituting the hydroxy-apatites.

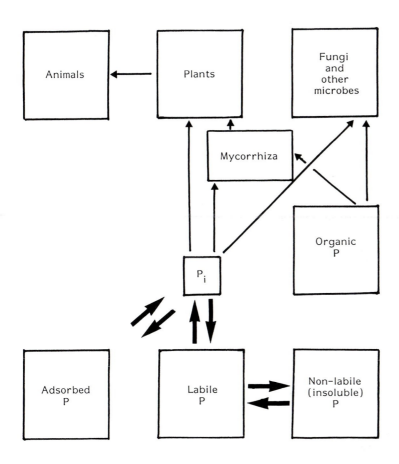

Figure 4.12: The various fractions in the soil and their relationships with living organisms. The sizes of the boxes are not significant, though the small size of that for orthophosphate in soil solution (P_i) is a reminder of the relatively small size of this phosphorus fraction compared with the others in the soil (the labile phosphate fraction is separate from the adsorbed but, in practice, the two may not be easily distinguished). Thick arrows between the inorganic soil phosphorus fractions indicate equilibria and the small size of the orthophosphate fraction is a result of them being in favour of the other fractions. The thin arrows give the direction of the *net* fluxes of phosphorus. Note the role of mycorrhizas in increasing phosphorus movement to plants (growing on phosphorus-poor soils).

The flowering plant, whether it be herb or tree, is only able to absorb orthophosphate; it cannot utilize organic phosphates. Further, a root, while being in contact with the soil solution and the phosphate within it, because of its size, will be only in restricted contact with the bulk of the labile pool associated with the soil particles. On the other hand a fungal hypha can potentially grow over these particles and have access to the labile pool as well as the soil solution. Furthermore, fungi can release phosphorus from organic phosphates through the action of phosphatases and bring into solution many otherwise insoluble phosphorus compounds.

The reliance of flowering plants on phosphate in the soil solution means that their growth is often limited under natural conditions by phosphorus supply. This is not only because much of the phosphorus in the soil is not readily accessible but also because phosphate can only move to the root by diffusion. Unlike nitrate, for instance, there is no movement of phosphate by the much faster process of mass flow, that movement of soil solution driven ultimately by transpiration from the leaves. The most effective way in which a flowering plant can increase the supply of phosphate is by extending its root system. In essence, such an extension increases the surface area capable of exploiting phosphate in the soil. The same capability is brought about by the hyphae from a root in the V-A mycorrhizal association (see *Figure 3.7b,* p. 41). When plants are in soils low in available phosphate, there is much better growth when the roots have formed a V-A mycorrhizal association than when the roots are uninfected.

Amongst mycorrhiza, the V-A are the commonest such association, in which an enormously wide variety of plants have roots infected by fungi belonging to the Endogonaceae. There is mycelium in the root which is characterized by the presence, in the cortical cells, of oval vesicles and arbuscules (meaning small tree), in which the hyphae branch repeatedly, the branches progressively decreasing in size, hence the term vesicular–arbuscular to describe the association. It is believed that the arbuscules form the interface between the fungus and root and that nutrient transfer takes place through them. The mycelium outside the root spreads extensively through the soil. In doing so, it extends the capability of the root to absorb phosphate and thus accounts for the better growth of mycorrhizal plants over those that are non-mycorrhizal. However, there is no evidence that the V-A fungi can break down organic phosphorus compounds.

For forest trees growing on acid soils, phosphorus is often not readily available. On the other hand, at leaf fall there is a potential flush of nutrients to the soil which could be available to the roots of the tree, provided they can compete with the microorganisms, including free-living fungi, which are also in the same habitat. The roots of such trees, as has been indicated earlier (see *Figure 3.7d,* p. 41), are covered with a mycelial sheath and such roots are found in the surface layers of the soil which comprises decomposed or partially decomposed leaves. In the case of the beech tree, *Fagus sylvatica,* it has been shown that when the sheath is present the flux of phosphate (moles per unit area per unit time) into such roots can be as much as five times that into roots which are not infected. It would seem that mycorrhizal roots have the enhanced capacity to absorb phosphate as it becomes available. Additionally, this type of mycorrhizal fungus is also able to hydrolyse organic phosphorus compounds. Thus the tree benefits from two fungal properties with respect to phosphorus acquisition.

The source of sulphur for fungi is either inorganic, the preferred form being sulphate, or low molecular weight organic molecules, such as the amino acid methionine. As yet there is no evidence for the hydrolysis of organic sulphur compounds.

Fungi are able to store phosphorus as condensed phosphate, polyphosphate, for the most part in the vacuole. For sulphur, the matter of storage is less clear but there is evidence from *S. cerevisiae* that glutathione can take on the function of a sulphur-storage compound.

4.4 Micro-elements

Carbon, nitrogen, phosphorus, sulphur, magnesium and potassium are considered as the macro-elements required by fungi for growth, in that proportionately more is required of these elements than of the other necessary elements. In the case of the first four, this is certainly so, because they make a major contribution to the structural components of the fungal cell or hypha. Magnesium or potassium do not have this role. The former is required as an activator of a number of key enzymes, while potassium is required to contribute to the appropriate ionic environment for enzyme functioning. Both these two elements, particularly potassium, make a significant contribution to the internal osmotic potential.

Those other elements required for growth at much lower concentrations are classified as micro-elements. They are calcium, iron, copper, manganese, zinc and molybdenum. It is not clear whether this is the complete list, since it is not easy to produce media of sufficient purity to ensure the absence of an element required at very low concentration. Also, and equally important, the requirement for growth and reproduction in the natural environment may not be the same as for vegetative growth in pure culture. Of the elements listed, the last five are known to be activators of enzymes, though in all probability this is a simplistic view of the role of these minor elements. If fungi behave like animal cells, these elements play a key role in total cell functioning and are subject to tight homeostasis.

Until recently, it was unclear as to whether or not calcium was essential for growth. It was known that presence of this element in the medium could, with different fungi, have a variety of effects including: growth, and sometimes morphogenesis, in some species of lower fungi; sporulation by some species of *Penicillium*; differentiation of the fruit body of *Chaetomium*; and branching in *Fusarium graminearum*. The present indications are that calcium is involved in cell signalling in a manner similar to its role in other eukaryotic cells. Commensurate with this role in fungi, one would anticipate a tight control of calcium concentrations in the cytoplasm (note what has been said above about the other micro-elements required by fungi), as part of the homeostatic mechanism acting on the element. Two components of such a mechanism have been identified in *N. crassa*, an outwardly directed (in relation to the cytoplasm) transport system at the plasmalemma and tonoplast, removing calcium from the cytoplasm into, respectively, the external medium and/or the vacuolar compartment.

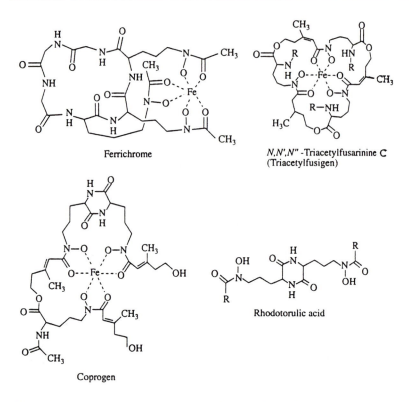

Figure 4.13: The structure of representative members of four hydroxamate siderophores produced by fungi. Reproduced from ref. 5 with permission from the British Mycological Society.

Although the micro-elements are required at low concentration, nevertheless in the natural environment these same elements may be in short supply. The ability of crop plants to show symptoms of deficiency of these same elements is testimony to their possible unavailability. Very often, unavailability is brought about by the element being made insoluble. Nevertheless, fungi have evolved mechanisms for scavenging the external medium for the necessary micro-elements. First, fungi possess transport systems which are not only highly specific for a particular micro-element but also possess a very high affinity for it. Second, fungi appear to produce compounds (ligands) which bind micro-elements and, when bound, the complex produced can itself be transported readily across the plasmalemma.

The striking example of the second mechanism is the production of siderophores. These are low molecular weight ferric-specific ligands which are involved in the solubilization and transport of iron in a whole range of bacteria and fungi (*Figure 4.13*). If the ferric ion is reduced, the ferrous ion shows little affinity for the ligand and this forms the basis for the removal of iron from the ligand. There are two ways in which siderophores can be involved in the uptake of iron by fungi. First, there is active transport of the whole iron chelate (Fe^{3+} siderophore) across the plasmalemma with release of the iron either by reduction or degradation of the siderophore molecule. Second, the siderophore acts solely as a molecule which captures iron and transports it to the plasmalemma, where the ferrous iron is released and transported across the plasmalemma via a permease.

Fungi produce a range of siderophores with a variety of binding groups for iron and differing overall molecular structure. Fungi also produce a range of peptides that can bind ions, including the other micro-elements. The ability to produce such metal ion-binding compounds indicates that micro-elements may not be readily available to fungi in many environments and the ability of a fungus to capture particular elements may put it at a competitive advantage. Thus, the possibility should not be discounted that competition for micro-elements is as, if not more, significant than competition for macro-elements. Certainly, there is good evidence that fluorescent pseudomonads can suppress *Fusarium* wilt and take-all disease by *Gaeumannomyces graminis* through the chelation of iron by siderophore molecules produced by the bacteria.

While required micro-elements are necessary at low concentrations for growth, they are frequently toxic at higher concentration. Elements such as copper and zinc are part of a group of what are colloquially called 'heavy metals', many with no known essential biological function, which are also toxic to living organisms. While heavy metals have always been found naturally in the environment, though in rather specific situations, they have become more widespread as a result of industrial activity. In spite of the potential toxicity of these elements, many fungi have the ability to cope in their close environments with the presence of many of them. Indeed *Aureobasidium pullulans* and *Cladosporium* spp. are fungi that are more tolerant of heavy metals and are better able than potential competitors to grow on oak leaves polluted with such metals.

Some of the mechanisms used by fungi to nullify the toxic effects of heavy metals are as follows.

(i)   *Extracellular complexation.* The best example is the production by the wood decay fungus *Poria placenta* of oxalate, which complexes copper and therefore allows the fungus to grow in the presence of copper-based wood preservatives. Oxalate is produced by many basidiomycete fungi. In woodlands, the large amount also produced complexes calcium. Indeed, in semi-arid regions it appears that fungal production of oxalate is important in the production of calcium sediments. There is evidence that complexation of calcium can occur in the walls of fungi in culture. While calcium is not a heavy metal, nevertheless, as indicated earlier, a fungus will need to maintain the appropriate cytoplasmic concentration. Complexation of calcium by oxalate can be seen as one way of counteracting a high external concentration. Fungi often produce large amounts of polysaccharidic material in culture. Most of the material has not been characterized, but experience with other microorganisms indicates that such polysaccharides may act as biosorbants of heavy metals.

(ii) *Processes at the plasmalemma.* In bacteria, there is good evidence that those species or strains which are more tolerant of heavy metals have decreased rates of uptake, either as a result of reduced influx of the metal or increased efflux. Some studies on *S. cerevisiae* suggest that similar mechanisms are likely to exist in fungi.

(iii) *Internal detoxification.* When potentially toxic metals are absorbed by living cells they are often made less harmful by compartmentation or incorporation into vacuoles, as has been demonstrated for calcium, magnesium and manganese. Certain fungi can be considered to detoxify the mercurous ion by reduction to the more volatile elemental mercury or methyl mercury.

4.5 Composting

Thus far, we have described fungal nutrition in somewhat simple terms. Here we indicate, through a consideration of composting, the situation that might apply in native soil, showing how the nutrition of a particular fungal species can be dependent on the activities of other fungi and microorganisms and how, as indicated in Chapter 3, the activities of fungi can alter the substrate on which they are living.

When vegetative material is collected together in a heap, fungal inocula are also incorporated: saprotrophic fungi from the plant surfaces such as roots and leaves; phytopathogens from the tissues of infected plants; fungi associated with soil particles; and spores transported there by many means, including by insects and small animals. The propagules produce mycelia and the mycelia already present develop further when the conditions are appropriate. Those fungi that develop first are those that consume small molecules released from the dying plant cells. The fungi concerned, zygomycetes and moulds, become visible as white mycelium and cause a colour change. This stage lasts only a few days. Other fungi attack the more resistant compounds, such as proteins and waxes. The most resistant compounds, such as keratin and lignin, are attacked only after several months when the material turns black and becomes more like the organic material in a humic soil.

For all this activity to take place, however, and also for the successful activities of earthworms and other animals, such as insect larvae, oxygen must be available. In the construction of the compost heap, it is therefore important to ensure proper aeration. The plant material itself, by giving a framework to the heap, allows channels for the penetration of oxygen, and additional plant material like straw with its central cavity can be very effective. Without oxygen, composting will cease.

Finally, within the compost heap there are other fungi as well as those degrading the plant material. From the outset there are mycophilous fungi, like *Syncephalastrum* and *Piptocephalis,* that attack other fungi, and there are also fungi that attack protozoa and nematodes. On the other hand, there are organisms other than filamentous fungi attacking the debris, like bacteria and yeasts which may be consumed by myxomycetes.

4.6 Metabolic regulation

When growing fungi in culture, one soon realizes that, as organisms, the great majority do not have very precise nutritional requirements. Most fungi can grow on a wide range of carbon and nitrogen sources and one imagines this must be the case for fungi growing in a compost heap. Thus we can think of fungi as being essentially omnivorous and because of this they can be opportunistic. However, omnivory makes biochemical demands on fungi. The evidence we have to date is that the pathways of primary metabolism in fungi do not differ very significantly from those of other eukaryotic organisms. Thus, within fungi, we know that glucose is metabolized via glycolysis or the pentose phosphate pathway to pyruvate which is oxidized via the citric acid cycle in mitochondria. Likewise nitrogen metabolism in its broad features is similar to that in the flowering plant. A fungus, however, is often able to utilize a wide variety of carbon sources and so there are many points at which carbon or nitrogen is fed into primary metabolism. Not all the entry points are capable of functioning at any one time. Depending on the substrate, enzymes are 'switched on or off'; in many instances the switching on requires *de novo* synthesis of the enzyme by the new nutrient entering the fungus. In this respect, the transport of the nutrients across the plasmalemma needs to be considered as part of the steps leading to the metabolism of the nutrient. Thus, not only may there be derepression of enzyme synthesis but the same can hold true for specific transport systems. Thus, in keeping with the development of a significant metabolic flux for the dissimilation of the nutrient, there has to be the development of the equivalent rate of supply of the nutrient from the medium to the protoplasm. Some of the points made above have been alluded to when considering the metabolism of nitrogen compounds by fungi (see p. 54).

Some of the above points can be illustrated by considering carbon metabolism. Fungi can use not only substances feeding directly into glycolysis such as glucose, fructose and sucrose but also compounds, such as glycerol, succinate, acetate, lactate and ethanol, which feed either into that part of metabolism at which pyruvate enters the citric acid cycle or into the cycle itself with probable involvement of the glyoxylate cycle (see p. 128). While energy for biosynthesis comes from the oxidation of these compounds via the citric acid cycle and the respiratory chain, there needs to be also a pathway to glucose such that, for instance, the polysaccharides of the wall may be synthesized. The glycolysis pathway is irreversible at those steps catalysed by hexokinase, phosphofructokinase and pyruvate kinase. Gluconeogenesis is the establishment of the pathway which bypasses these three irreversible steps. When *S. cerevisiae* is growing on ethanol, both phosphofructokinase and pyruvate kinase are repressed, while fructose-1,6-bisphosphatase (which allows fructose-1,6-bisphosphate to be metabolized to fructose) is derepressed.

Another example of metabolic regulation relates to changes at the organizational level. The great majority of fungi cannot grow on compounds that contain only a single

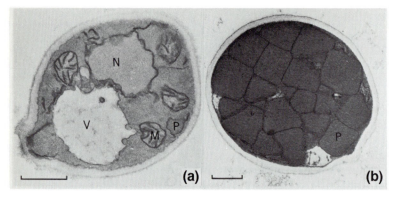

Figure 4.14:   Electron micrographs of sections of cells of the yeast *Hansenula polymorpha*, showing how the internal organization is altered from when (a) cells are growing on glucose to (b) when cells are growing on methanol. n, nucleus; v, vacuole; p, peroxisomes; m, mitochondrion. Scale marker represents 0.5 μm. Photograph by M. Veenhuis.

carbon atom such as methanol. Indeed before 1969, it was believed that such one-carbon compounds could not be used by fungi. Now a number of yeasts, such as *Candida boidinii*, have been shown to grow on methanol and the biochemistry of the process is well established. The essentials of the process are the oxidation by molecular oxygen of methanol to formaldehyde:

$$CH_3OH + O_2 \rightarrow HCHO + H_2O_2.$$

This is followed by the complete oxidation of formaldehyde to carbon dioxide with the concurrent formation of reduced nicotinamide adenine dinucleotide (NAD). The reduced NAD provides the free energy for the synthesis of glyceraldehyde-3-phosphate from formaldehyde. There is, of course, a need to remove the hydrogen peroxide produced in the first oxidation step. This is brought about by catalase. This enzyme is located in an organelle, the peroxisome. When the yeast cells grow on methanol, the number of peroxisomes increases very substantially so that, instead of there being a few in the cell and of small volume, there can be many of large volume (*Figure 4.14*). Thus there is not only a change in the enzyme complement of the yeast cell when transferred from glucose to methanol (enzymes responsible for the oxidation of the latter are repressed by the hexoses) but there is a striking change in the subcellular organization.

The metabolism of some fungi can be considered to be unbalanced, in the sense that they produce large quantities of primary metabolites (or compounds closely related to such metabolites) which are found in the external medium. A number of these compounds are of commercial importance and using fungi may be the most effective means of producing them (*Table 4.2*). Yields have been increased by strain selection and manipulation of the composition of the culture medium. In the case of citric acid production by *Aspergillus niger*, yields have been increased by using a manganese-deficient medium.

Table 4.2:   Primary metabolites produced industrially from fungi

Metabolite	Fungal species used
Ethanol	*Saccharomyces cerevisiae*
Citric acid	*Aspergillus niger*
Itaconic acid (methylene succinic acid)	*Aspergillus terreus*
Gluconic acid	*Aspergillus niger*
Fumaric acid	*Rhizopus arhizus*

4.7 Novel substrates

This chapter would not be complete without a reference to the fact that fungi can grow on substrates not found in nature. A striking example, a problem of considerable applied importance, is the ability of fungi, particularly *Cladosporium resinae*, to grow in aviation kerosene, causing blocked fuel lines, as well as initiating metal corrosion. There are many industrial products which would be susceptible to fungal attack were it not for the inclusion of biocides in the formulation or production of these products.

References

1. Adler, E. (1968) *Svensk Kem. Tidsk.,* **80,** 279–290.
2. Leisola, M.S.A. and Garcia, S. (1989) in *Enzyme Systems for Lignin Degradation* (M.P. Coughlan, ed.). Elsevier Applied Science, London, pp. 88–89.
3. Read, D.J. (1984) in *The Physiology and Ecology of the Fungal Mycelium* (D.H. Jennings and A.D.M. Rayner, eds). Cambridge University Press, Cambridge, pp. 213–240.
4. Smith, D.C. and Douglas, A.E. (1987) *The Biology of Symbiosis.* Edward Arnold, London.
5. Winkelmann, G. (1992) *Mycol. Res.,* **96,** 529–534.

Chapter 5

Water: living with desiccation

5.1 Introduction

As with all living cells, fungi need water to grow. As we have seen in Chapter 1, turgor is the hydrostatic driving force for hyphal growth just as it is in green plant cells. There are situations where it is necessary to conserve water. Thus fungi produce thick-walled spores and there are multicellular organs, such as fruit bodies, rhizomorphs and sclerotia that have thick-walled outer layers. One presumes that the thick walls are all adaptations to reduce water loss, but there are fungi that can grow or survive in very dry environments without such morphological adaptations. A dry environment should not be considered as just being without water: the water may be present but not readily accessible. Such inaccessibility is mostly due to the presence of dissolved substances, giving an unfavourable osmotic value to the medium and making it difficult for an organism in it to absorb water.

Here we describe three situations where fungi grow in spite of apparent problems of water supply. The first two situations (marine fungi and xerophilic fungi) are similar in that the fungus has adapted to growth in a medium of unfavourable osmotic value. However, the marine fungi have their own intrinsic interest since they occupy a very important ecological situation and, in their evolution, have become adapted, not only to the more unfavourable osmotic value of seawater compared to terrestrial aqueous systems, but to other aspects of their environment. Therefore marine fungi have been treated here separately from xerophilic fungi. The third situation is represented by lichens. These associations are of particular interest because they are frequently exposed to cycles of wetting and drying, which poses considerable problems for the organisms if they are to remain viable.

5.2 Marine fungi

There is much combined carbon in the sea. Within the oceans, it is a consequence of the death of marine organisms. This is also true of the shoreline, particularly those harbouring mangrove and macro-algal (seaweed) communities. Other forms of organic carbon are excreta, slimes and mucilages, and sexual and other attractants from marine organisms. Additionally, there is import of combined carbon from rivers and there is considerable bulk of non-living combined carbon in the timber constructions made by man. Finally, we must not forget the large amount of crude and refined oils spilling into the sea. All of this combined carbon forms potential substrates for fungi.

Many collections of marine fungi have been made from all over the world, from a wide diversity of habitats. We now know that the marine fungal flora is very diverse, containing many species from the lower fungi, ascomycetes and deuteromycetes *(Table 5.1)*. Surprisingly, there are only a few basidiomycetes that are marine.

Although the great majority of marine fungi differ morphologically (in their asexual spores and sexual reproductive structures) from terrestrial fungi, it has been argued that a significant number are secondarily marine. Certainly, a considerable number of what can be legitimately termed terrestrial fungi are capable of living in marine habitats. On the other hand, it is believed that quite a number of species have originated from marine algal ancestors. A small number of species amongst the 'lower fungi', but of otherwise uncertain affinities, clearly belong to an independent group. Indeed it is not certain whether or not they are fungi. They require high concentrations of sodium for growth and they possess an unusual cell organelle and an outer covering of scales of unusual composition for fungi (sulphated polysaccharide).

The evidence for certain marine fungi having evolved from marine algal ancestors is essentially morphological. There is clearly a case here for nucleic acid analysis of appropriately chosen fungal species and their putative algal relatives. We have

Table 5.1: Semi-quantitative measure of the number of species of marine fungi known to colonize different substrates in the sea[a]

Substrate	'Lower fungi'	Ascomycetes	Basidiomycetes	Fungi imperfecti
Wood	+	++++	+	++
Free floating	+	+++	++	++
Sediments, mud, soil, sand dunes	+	++	−	++++
Algae	+++	+++	+	++
Animals	++.	+	−	+
Decaying leaves of terrestrial plants	+	+	−	++
Decaying leaves of marine angiosperms	++	+	−	+++
Mangrove plants	+	++++	+	++

[a] Key to symbols: −, none known; +, 1–5 species; ++, less than 10 species; +++, less than 50 species; ++++, more than 50 species. This table is based on information supplied by E.B. Gareth Jones.

pointed out that the other part of the marine fungal flora probably arose from invasion of the sea by terrestrial species. Although some species are common to both terrestrial and marine habitats (and we need to remember here that isolation of a species from the sea does not necessarily mean that the particular species is able to complete its life cycle in the sea), there is no doubt that the marine fungal flora is overwhelmingly different from that which might be isolated from a terrestrial habitat. This shows that the marine environment exerts special physiological pressures on the functioning and development of fungi.

Compared with a terrestrial environment, the sea has the following features that make specific physiological demands on fungi. First, because of the high concentration of salts, a marine fungus must generate a much higher osmotic pressure than a terrestrial fungus to produce the requisite turgor for hyphal or cell growth. Second, a major component of the sea is sodium chloride which, if absorbed to generate the necessary osmotic pressure, would need to be, from our knowledge of flowering plant halophytes, sequestered in large vacuoles because of the toxicity to cellular mechanisms of sodium and chloride ions when they are at high concentration. Indeed it has been shown *in vitro* that sodium chloride at the concentration in seawater can be inhibitory to fungal enzyme activity.

The evidence to date is that marine fungi in fact exclude sodium chloride, not totally, but sufficiently to keep the two constituent ions at a concentration in the protoplasm such that they do not have much, if any, inhibitory effect on metabolism. There is no preferential accumulation of sodium chloride in the vacuoles. In fact, unlike flowering plants, where the vacuole in the fully grown vegetative cell is 90% of its volume, in marine fungi, vacuoles constitute no more than about 20% of the volume of the protoplasm. The shortfall in osmotically active solutes brought about by the restriction in the absorption of sodium chloride is made up by other inorganic ions such as potassium and calcium (partially balanced by organic acids synthesized by the fungus) but principally by the absorption or synthesis of organic compounds. These organic compounds (which increase in concentration not only within marine fungi but also in other fungi in response to osmotic stress) are polyols, especially in the eumycota *(Table 5.2),* and amino acids, particularly proline, in lower fungi, including those marine species that require sodium for growth.

The major polyhydric alcohol (polyol), which is synthesized under saline conditions, is the three carbon glycerol, but there are other polyols produced by fungi, namely the six carbon mannitol and the five carbon arabitol, both of which can make a significant contribution to the osmotic pressure of a fungus (see *Table 1* in the Introduction). Both are found in marine fungi. These polyols, as well as the amino acid proline and the disaccharide trehalose, which is found in many fungi but not in those which are marine, have an important property in common, that is, that at high concentration (1 M or more) they have no significant effect on enzyme activity. Thus, unlike salt solutions, they do not affect protein conformation. Because of this property, they are called *compatible solutes.*

Table 5.2: Principle soluble carbohydrates that can be found to a greater or lesser extent in the asco- and basidiomycetes

Polyhydric alcohols (polyols)

Trivial names	Glycerol	meso-Erythritol Erythritol[a] Erythrol	D-Arabinitol D-Arabitol[a]	L-Arabinitol L-Arabitol[a]	Xylitol[b] –	D-Glucitol Sorbitol[a]	D-Mannitol

Formula:

Glycerol:
$$\begin{array}{c} CH_2OH \\ | \\ HCOH \\ | \\ CH_2OH \end{array}$$

meso-Erythritol:
$$\begin{array}{c} CH_2OH \\ | \\ HCOH \\ | \\ HCOH \\ | \\ CH_2OH \end{array}$$

D-Arabinitol:
$$\begin{array}{c} CH_2OH \\ | \\ HOCH \\ | \\ HCOH \\ | \\ HCOH \\ | \\ CH_2OH \end{array}$$

L-Arabinitol:
$$\begin{array}{c} CH_2OH \\ | \\ HCOH \\ | \\ HOCH \\ | \\ HOCH \\ | \\ CH_2OH \end{array}$$

Xylitol:
$$\begin{array}{c} CH_2OH \\ | \\ HCOH \\ | \\ HOCH \\ | \\ HCOH \\ | \\ CH_2OH \end{array}$$

D-Glucitol:
$$\begin{array}{c} CH_2OH \\ | \\ HCOH \\ | \\ HOCH \\ | \\ HCOH \\ | \\ HCOH \\ | \\ CH_2OH \end{array}$$

D-Mannitol:
$$\begin{array}{c} CH_2OH \\ | \\ HOCH \\ | \\ HOCH \\ | \\ HCOH \\ | \\ HCOH \\ | \\ CH_2OH \end{array}$$

Disaccharide

	α-D-glucopyranosyl- α-D-glucopyranoside
Trivial name	Trehalose[a]

Formula

---

[a] Name in common usage.
[b] When fungus grows on D-xylose.

In terrestrial fungi, those processes relating both to asexual spore production and sexual reproduction appear to be especially sensitive to sodium chloride. Although we have absolutely no idea of the underlying physiology, marine fungi have clearly evolved reproductive mechanisms adapted to life in the sea. Morphological adaptations are very evident; for instance, the asexual spores and the sexually produced ascospores frequently have appendages to give the spores increased flotation and thus better capability for dispersal *(Figure 5.1)*. The appendages, when in contact with the substratum, often generate mucilage-like material which appears to act as an adhesive. When the spores germinate, the emergent hyphae also produce mucilage to anchor them to the substratum *(Figure 5.2)*. Similarly, ascomycete fruit bodies have adapted to the sea by releasing spores through the swelling of mucilage, since the terrestrial mechanisms by which the spores are shot out of the fruit body are no longer possible.

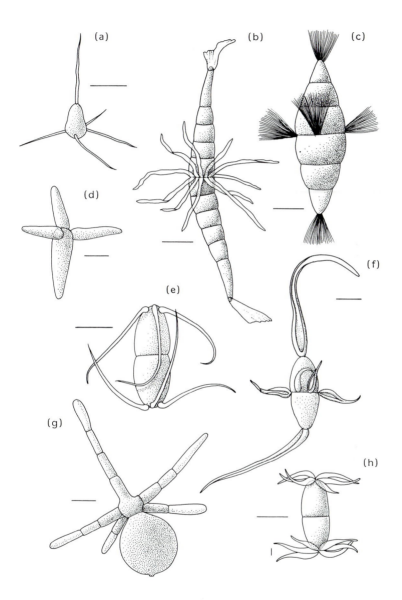

Figure 5.1:   Some appendaged spores of marine fungi. (a) *Nia vibrisa;* (b) *Corollospora pseudopulchella;* (c) *Nereiospora comata;* (d) *Digitatispora marina;* (e) *Arenariomyces trifurcatus;* (f) *Octaspora apilongissima;* (g) *Orbimyces spectabilis;* (h) *Remispora stellata.* Bar represents 10 μm. Redrawn from an original by J. Kohlmeyer and B. Volkmann-Kohlmeyer.

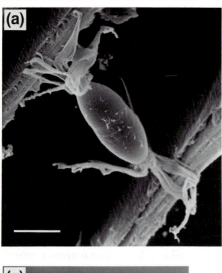

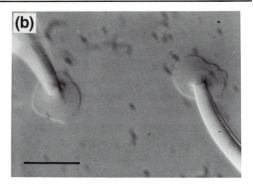

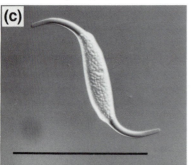

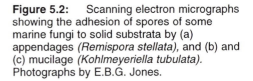

Figure 5.2:   Scanning electron micrographs showing the adhesion of spores of some marine fungi to solid substrata by (a) appendages *(Remispora stellata),* and (b) and (c) mucilage *(Kohlmeyeriella tubulata).* Photographs by E.B.G. Jones.

5.3 Xerophilic fungi

It is well known that food can be preserved by keeping it in concentrated salt or sugar solutions. Salted fish and meats, and jams and other fruit-based products are testimony to the success of this preservation procedure. However, there are fungi that can invade these environments, which are totally hostile to bacteria. The fungi that do so are called xerophilic fungi, since the feature common to all environments inhibitory to the great majority of microorganisms, is the extremely low osmotic potential of the medium. Furthermore, these fungi show the unusual property of growing much better in concentrated (but less concentrated than the maximum tolerated) salt and sugar solutions than in solutions that are dilute. We do not know why this is so. Included in the category of xerophilic fungi are moulds, such as *Aspergillus (Eurotium)* and *Penicillium* species, and the yeasts *Debaryomyces hansenii* and *Zygosaccharomyces rouxii,* which grow in solutions of sodium chloride around 5–6 M. Some xerophiles can grow in solutions that are even more concentrated, namely *Chrysosporium fastidum, Monascus bisporus,* some isolates of

Aspergillus restrictus and *Zygosaccharomyces rouxii,* when the external medium is sugar not salt. Xerophiles generate the appropriate osmotic pressure (and therefore the required turgor) within their protoplasm by synthesizing glycerol or absorbing a solute from the medium which is compatible with the metabolic machinery.

5.4 Lichens

These associations of fungi with an alga or cyanobacterium, unlike almost certainly the majority of fungi, can function or survive in environments where the water content varies considerably. As indicated in Section 5.1, lichen thalli are able to withstand repeated wetting and drying. A dry lichen is without detectable physiological activity and can survive long periods of desiccation. It is believed the polyols (mannitol, sorbitol, arabitol and ribitol) produced by the fungus, and sometimes by the alga, act as protective agents by replacing that water associated with macromolecules. Trehalose appears to play a similar role in fungal spores, which are also structures with very low water content.

On rewetting of a lichen, there is a rapid rise in respiration and a marked capacity to lose solutes, polyols and inorganic ions into the surrounding medium. As hydration continues, the membranes increase their water content, turning from a form where the phospholipid is present as discrete micelles to a continuous bilayer, now impermeable to small molecules and ions except via the protein carrier molecules (see *Figure 13.2*). As membrane integrity is re-established, respiration returns to its basal rate while photosynthesis rises. The time at which metabolism returns to normal depends on the lichen. In the desert, this process of dehydration and rehydration occurs in the day and in the night respectively, the water in the latter phase coming from the dew which is deposited in the dark. Consequently, a crucial period for the lichen is the period after sunrise when photosynthesis takes place before water loss inhibits metabolism.

Chapter 6

Oxygen and lack of it

6.1 Introduction

We have seen that oxygen must be available for hyphal extension. The generation of the precursors and the maintenance of polarity at the hyphal tip requires metabolic energy provided by the respiratory oxidation of carbon compounds. The role of oxygen influences all stages of development and differentiation. On the other hand, fungi often live in environments with low or zero oxygen tensions, as in deeper layers of the soil or inside large trunks or heaps of compost or dung. These microaerobic and anaerobic environments are not typical domains for fungi and justify special treatment.

6.2 Life in air at 'normal' oxygen tensions

During the vegetative phase, the energy requirement, and therefore the oxygen requirement, can be comparatively small. This is not entirely surprising, since, unlike a green plant, which synthesizes its carbon compounds from carbon dioxide, the starting point for the synthesis of polymeric compounds in a fungus is frequently at the level of hexose and amino acid, which have entered directly from the external medium.

However, it is relatively easy to identify situations in which fungi have increased oxygen requirements. We have already instanced the degradation of lignin as a process requiring the significant intervention of oxygen. A similar kind of process is the oxidative polymerization of phenols initiated by enzymes called laccases, as shown in *Figure 6.1* with the aromatic amino acid phenylalanine as the starting point. A variety of compounds can be metabolized in this manner; some of them are toxic to fungi, so the process can be regarded as a detoxification process.

During the process of polymerization illustrated in *Figure 6.1*, it can be seen that there is an increasing formation of conjugated double bonds. This increase leads,

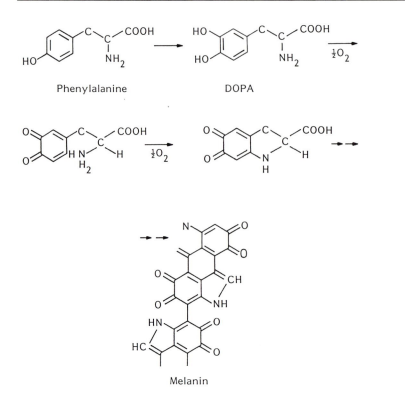

Figure 6.1: Synthesis of melanin from phenylalanine via oxygen-requiring steps to 3,4-dihydroxyphenylalanine (DOPA) and related compounds, and then polymerization reactions.

within the compounds synthesized, to an increasing ability to absorb light, resulting in the compound being coloured and as more double bonds are formed the compound becomes black. Such colour changes can be seen when mycelia age. Old mycelia or fruit bodies (perithecia) of ascomycetes become green and later black. Some mushrooms turn red or blue when cut, that is, when exposed to oxygen. This reaction can be rapid, as in *Boletus luridus* (see cover).

The type of reactions just described play an important role in the formation of soil and in composting, where aromatic compounds like tannins or those resulting from the breakdown of lignin are recondensed non-enzymically, along with other nitrogen-containing compounds, into humates (humic acid, fulvic acid) with a molecular weight of up to 30 kDa. They are fixed on to soil particles or clay minerals and they themselves absorb cations, amino acids or anions, binding them with ionic, covalent or van der Waals forces. A suggested structure is given in *Figure 6.2*. There is an important point to stress here. Whilst we realize that much of the activity of a fungus

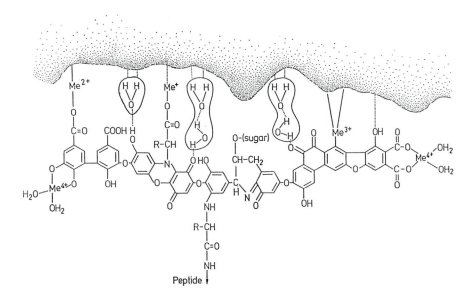

Figure 6.2: A hypothetical humate molecule bound to a clay particle (stippled area) showing the possible ionic and covalent (—) and van der Waals (...) bonds between the molecule and the particle. Me, metal ion.

in its environment is degradative, we must not ignore the capability of fungi to synthesize or initiate the synthesis of complex organic molecules. Some of these molecules come under the category of secondary metabolites (see Section 10.2, p. 101). However, whichever type of molecule is produced, it will influence the niche in which the fungus lives – in the case of humates it is clear that the influence can be profound.

With the onset of differentiation there is an increased oxygen demand. Increased consumption of oxygen has often been demonstrated when hyphal growth ceases and reproduction is initiated. When the formation of propagules (spores, etc.) and other processes of differentiation take place, there is a far higher requirement for protein and nucleic acid synthesis than in the case of hyphal extension, in which there is much greater emphasis on the synthesis of wall material and, to a lesser extent, lipid. Energetically, protein and nucleic acid synthesis make much higher demands on the fungal metabolism, explaining the increased oxygen utilization. This means that differentiation can only take place where this increased demand for oxygen can be fulfilled. So there is an additional reason, as well as access to air currents for spore dispersal, why most fungi form their fruit bodies at or near the surface of their substrates where the necessary oxygen demand can be fulfilled.

6.3 Microaerobic life

As indicated above, the fungal mycelium often lives and grows within the substrate. Here the available oxygen is limited, since the rate of diffusion from the ambient air at the surface to the hyphae will not be sufficient to maintain a growth rate anything like that of hyphae if they were on the surface. This is clearly seen in static cultures in the laboratory (see p. 19). However, microaerobic environments are still accessible to filamentous fungi, with consequent degradation of the substrate. But fungal growth and substrate degradation occur at a very slow rate. This is exemplified by doubling times of very long duration, with doubling times inside substrates like soil or wood of up to a year. Of course, limitation of nutrients may also be a contributory factor in reducing the growth rate.

6.4 Anaerobiosis: the yeasts

Total lack of oxygen prevents fungal growth, except for some very special fungi which are found in sites with no or very low oxygen tensions, such as sewage, silage or the rumen of cattle or sheep. Fungi isolated from the rumen, such as *Neocallimastix frontalis,* are strict anaerobes, possessing no mitochondria, and are very effective in breaking down cellulose.

The rumen is a nutritionally rich habitat in which continuous anaerobic conditions are brought about by the oxygen demand of the large number of microorganisms present. There are similar substrates where the same situation can occur, such as nectar in flowers, dung balls and fleshy fruits. The little oxygen that is present initially is consumed rapidly. Under these circumstances there is a switch to fermentation as a source of metabolic energy.

This fermentative metabolism, or at least the capability for it, is characterized by the formation of single cells, which might form chains but have no cytoplasmic contact with each other. Yeasts, as this growth type is called, divide by fission or budding and, unlike hyphae, are not capable of penetrating far into a particular substrate. Their liquid habitat provides them with the necessary nutrients at concentrations which are ample for growth. Metabolically, the yeast state is characterized by fermentation, the production from the nutrients consumed of the necessary metabolic energy and precursors for growth without the intervention of molecular oxygen (though it should be remembered that fermentation can occur under aerobic conditions, though at oxygen tensions lower than normal air). Fermentation in its most common form is known from baker's yeast, which transforms a variety of sugars to carbon dioxide and ethanol via glycolysis, which is the source of metabolic energy. Another product of fermentation under anaerobic conditions is lactic acid but essentially this compound is produced in oomycetes, chytridiomycetes and zygomycetes.

In contrast to obligately anaerobic fungi and bacteria, in which there has been a total switch to fermentation, as soon as air becomes available fungi switch back to oxidative metabolism, which is more energetically productive than fermentation. Indeed, oxygen may be necessary for continued growth, because fermentation cannot underpin the synthesis of certain compounds, which requires molecular oxygen, if they are unavailable in the external medium.

The metabolic switch between aerobic and fermentative metabolism can be loosely paralleled by morphological changes, as is seen in some species of *Mucor*. In an aerobic environment with moderate nutrient supply they grow in a filamentous manner; with decreasing amounts of oxygen and sufficient available nutrients they gradually alter their form to the yeast-like type, as is shown in *Figure 6.3*. Similar morphological switches also occur in other fungi, though the trigger for the switch has not been properly identified.

The yeast form has another important property, that is, not only is it the unit of vegetative growth but it can also act as a dispersive propagule; for instance, yeasts growing on nectar will be dispersed by pollinating insects to sources of nectar as yet uncolonized. Thus the yeast cell is here strictly analogous to a spore.

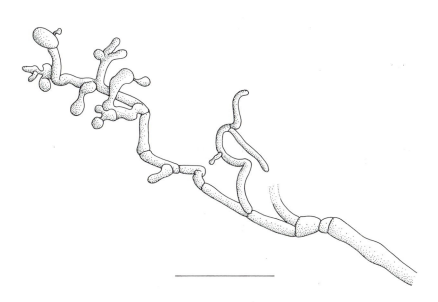

Figure 6.3:   Transition from filamentous to yeast-like growth in *Mucor hiemalis* brought about by the decreasing availability of oxygen in the deeper layers of a malt-extract medium. Bar represents 50 µm. Drawing by kind permission of G. Duvinage.

Chapter 7

Using light

7.1 Introduction

One would not anticipate that fungi might have a requirement for light. They are grown routinely in the laboratory in the dark and in nature are often found growing in the absence of light. However it will be shown here that fungi have developed sophisticated relationships with the daily light–dark cycles, particularly with respect to the initiation of reproduction.

7.2 The action of light

It is the generally accepted view, which is experimentally well-supported, that for fungi light in the short wavelengths, blue light, is effective, while light with longer wavelengths is ineffective. Red light is without effect and so can be used as a 'safe light' in experiments in which it is necessary to inoculate and grow fungi in culture under conditions which for them is dark. It was inferred from the known action spectrum *(Figure 7.1),* and later confirmed by biochemical means, that riboflavin is the light receptor. The required quantities of light are low, well below those of the full moon on a clear night. Examples of the effects of light on fungi, involving all groups, are given in *Table 7.1.*

Thus the reaction chain from the photons hitting the fungus starts with their absorption by riboflavin, which then reduces a *b*-type cytochrome, as has been found to be the case in *N. crassa*. Little is known about how the recycling process, or reoxidation in the dark, affects hyphal metabolism. It is evident that ionic movement in the hyphal tip region is affected, such that there is interference with extension.

The primary events of irradiation seem to be similar in most fungi. While the entire chain of events following the exposure to blue light is not known, the consequences of irradiation soon become evident. With *Trichoderma* in culture, light leads to staling followed by formation of conidia. Translating this response to nature, light is the

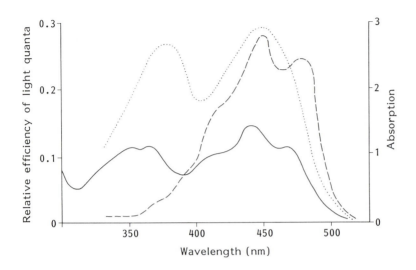

Figure 7.1: The absorption spectra of riboflavin (....) and β-carotene (---) in hexane and the action spectrum for the phototrophic response of the sporangiophore of *Pilobolus* (see *Figure 7.6*).

signal that the mycelium has reached the (irradiated) surface, where there can be production of spores in an environment accessible to those processes responsible for the production and transport of the spores. In other instances, light overcomes developmental blocks, as has been shown for *Coprinus comatus*, in which fruit body primordia do not develop further without light.

Table 7.1: Representative responses of fungi to low levels of light

Organisms	Responses	Light levels
Coprinus lagopus	Increased fruiting	0.1 lumen m^{-2} for 5 sec
Penicillium isariiforme	Sporulation	0.85 μW cm^{-2}
Sclerotinia fructicola	Growth (maximum rate) of conidiophores	0.1–0.3 μW cm^{-2}
Phycomyces blakesleeanus	Initiation of sporangiophores	6 x 10^{-4} μW cm^{-2} (saturation level; some reponse occurs at <7 x 10^{-5} μW cm^{-2})
P. blakesleeanus	Bending of sporangiophores	0.04 μW cm^{-2} (min. of normal range; total range extends to 3 x 10^{-7} μW cm^{-2})
Neurospora crassa	Suppression of circadian rhythm	0.02 μW cm^{-2}
Moonlight	(Full; clear sky)	0.23 μW cm^{-2}

Reproduced from ref. 1 with permission from Academic Press.

7.3 Light–dark cycles

We know little about the role of light in the development of fungi living in their natural habitats. However, in both nature and experiments, often it is not light itself that triggers the response but the onset of light after a dark period. Since this is so, the daily dark–light change has the same effect each day. Thus structures induced by light are formed repeatedly. With the bulk of the mycelium growing mainly within the substrate, that reaching the surface is stimulated each morning to differentiate. When the exposed hyphae are irradiated, their rate of extension is reduced, preliminary to differentiation into reproductive structures. Those not exposed continue to grow trophically at the original rate and eventually fill that space that would have been filled by those hyphae that have been irradiated and whose trophic growth has been terminated. Since the mycelium spreads more or less in a circular fashion from the point of origin, the reproductive structures are located in concentric rings around the origin. These rings are well known from the ring-rot of apples caused by *Sclerotinia fructigena* (*Figures 7.2* and *7.3*) but also from other fungi.

This process of rhythmic growth and reproduction is important ecologically. The process brings about propagation (via spore dispersal) at an earlier stage than would be the case if propagation were dependent on reaching the edge of the substrate or an equivalent barrier to further growth. That only selected (those on the surface of the substrate) hyphae are stimulated to differentiate into reproductive structures, allows the bulk of the mycelium to continue growing. As one can see in *Figure 7.4,* which gives results from an experiment with *Trichoderma,* the ability of the mycelium of a rhythmically growing and conidiating strain to produce spores throughout most of its

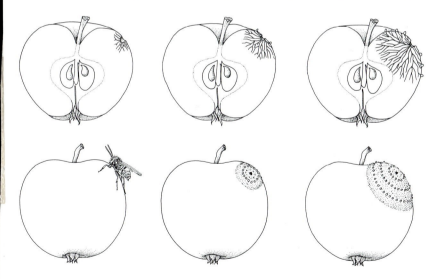

Figure 7.2: Diagram of the development of rings of clustered conidiophores *Sclerotinia fructigena* as shown in *Figure 7.3*.

Figure 7.3:   Photograph showing ring-rot caused by *S. fructigena* on an infected apple.

growth period enhances markedly the number of spores produced over the number produced by the uniformly growing strain and therefore improves the reproductive success of the former strain.

Rhythmic fungal growth is very familiar in nature as 'fairy rings' (*Figure 7.5*). It has been found that in certain regions most basidiomycetes form such rings. As we indicated in the previous paragraph, whenever a fungus is growing in a substrate that is large in relation to the size of the colony, fairy ring formation improves the reproductive success of the colony, yet allowing the mycelial extension for the acquisition of new substrate and nutrients.

The periodic behaviour of the kind described above, which is produced exogenously as the result of an alternation of an environmental variable such as light, is termed diurnal rhythm. However it can also be endogenous, as is the case in *Sclerotinia fructigena*. Thus there is a link between fungal rhythms and the biological clock found in plants, animals and man, based on circadian rhythms. Such rhythms are exceptional in fungi. They have been demonstrated in the (experimentally induced) mutants and specially manipulated cultures of *N. crassa*. Nevertheless, it is doubtful whether circadian rhythms occur in fungi in nature.

7.4 Using the direction of irradiation

Many fungi also use the direction from which the light is coming to orientate themselves. Thus the conidiophores or sporangiophores of many fungi grow in the

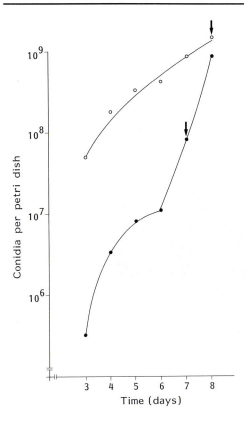

Conidia per petri dish

Time (days)

Figure 7.4: Spore production in a rhythmically growing and conidiating strain of *Trichoderma* sp. (o) and a uniformly growing (●) strain. The arrows mark the time when the growing mycelium reached the edge of the petri dish.

direction of the light source, most obviously so in the large sporangiophores of *Phycomyces*. Even more spectacular is the active shooting off of the spores or sporangia into the same direction. This is most obvious with *Pilobolus,* where the process has been studied in detail. In *Figure 7.6* it can be seen that the sporangiophore has a large subsporangial bladder in which a hydrostatic pressure is built up, water being provided by the 'root bladder', which is in contact with the substratum. Below the subsporangial bladder there is a ring of yellow-orange carotenoids, which is shaded by the dark mass of spores in the sporangium. If the light received by the ring is at a minimum, that is, the ring is in the darkest part of the shadow, the axis from the ring through the middle of the sporangium, the longitudinal axis of the subsporangial bladder, gives both the direction of the light and the direction in which the sporangium will be shot off. If the light received by the ring is not at a minimum, the sporangial stalk bends until it is. The mechanism is very subtle, for from a distance of 40 cm the sporangia can hit, in an otherwise dark box, a glass window of 5 cm diameter.

Figure 7.5:   Fairy ring of fruit bodies of *Clitocybe* sp.

7.5 Growing in the light

The 'typical fungus' described in earlier chapters lives and grows within the substrate and thus in the dark. Light, even in low quantities, indicates to such fungi the surface of the substrate. Some fungi can grow permanently on sites exposed to light and thus avoid competition with the bulk of photosensitive species, though the physiological basis for the light insensitivity has not been explored. Two ecological groups have invaded this niche, fungi growing on exposed surfaces and those plant-parasitic fungi infecting their hosts in the above-ground parts. We shall only deal with the first group here; the second group is considered in a general way in Chapter 3 (see p. 32).

Plant surfaces, of which leaves have been the most extensively studied, bear a rich fungal flora growing and sporulating directly on the surfaces concerned. The most common species are listed in *Table 7.2*. The pigmentation of these fungi, at the least their spores are pigmented, is interpreted as a protection against the heavy irradiance. The list contains species that are also found in lists of air spora, not surprising in view of the fact that the spores are liberated from surfaces (to colonize other leaf surfaces) well into the turbulent layers of the atmosphere.

There are plenty of nutrients available for the growth of these surface-borne fungi: dust sedimented from the air, exudates from the leaf, excrement and exuviae of

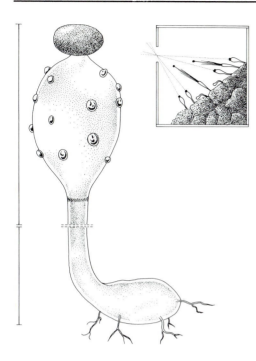

Figure 7.6: The sporangiophore of *Pilobolus* showing: the apical sporangium, with the preformed zone of rupture below; the subsporangial bladder, whose internal hydrostatic pressure is increasing (note the water droplets on the surface); the ring of carotene-rich protoplasm below the bladder; the sporangiophore with the basal bladder attached to the substrate. Inset: the experimental system to show how sporangiophores are aimed in the direction of light. Scale marker represents 10 mm.

arthropods, sap, originally from the phloem, exuded from aphids (particularly rich in sugars and which on lime or oak leaves can support the growth of so many fungi that the leaves turn black). Equally, fungi on leaves are in an exposed position to any toxic compounds that land on the leaf. Thus in urban areas the number and composition of the leaf-surface fungal flora are affected by sulphur dioxide in a manner that allows the observed changes to act as a measure of the concentration of the gas to which the leaves have been exposed.

The environment on the leaf surface and the fungi which live there is called the phylloplane. There appears to be a smooth transition from the saprotrophic fungi living there to the other fungi living in such irradiated sites, plant parasites. Many of the phylloplane fungi are potential parasites, for example, *Botrytis cinerea*, *Cladosporium* spp. or *Alternaria* spp. These fungi also grow saprotrophically but they invade the plant as soon as the conditions are suitable. This is the case if the host is under stress. Infection also occurs at the end of the vegetative phase, when leaves become senescent.

On the other hand, typical phylloplane fungi compete with one another and thus with the potential parasites during the saprotrophic phase. Thus, if the population of the harmless phylloplane fungi like *Aureobasidium pullulans* or *Sporobolomyces* is dense enough, the growth of potential parasites is restricted and thus infection is reduced.

Table 7.2:   Microscopic fungi found regularly on the
surfaces of leaves

Alternaria alternata
Aureobasidium pullulans
Botrytis cinerea
Cladosporium cladosporioides
Cladosporium herbarum
Cladosporium sphaerospermum
Epicoccum purpurascens
Exophiala jeanselmei
Phoma pomorum
Sporobolomyces roseus
Trispospermum pinophilum
Trispospermum sp.

However, attempts to use these saprotrophic phylloplane fungi (mainly *A. pullulans*) as agents for biological control has been only partially successful. To maintain the required high densities it is necessary to spray spores of the biocontrol fungus on to the leaf frequently; otherwise the population returns to its original size and composition.

Reference

1.   Tansey, M.R. and Jack, M.A. (1975) *J. Theoret. Biol.*, **51,** 403–407.

Chapter 8

Withstanding extremes of temperature

8.1 Introduction

Fungi have a range of temperatures in which they are able to live. These are what are termed the 'physiological' temperatures, at which mycelial growth and reproduction are possible and the range will be somewhere between 0 and 40°C. There is a wider range in terms of survival of fungi as resting or dehydrated structures. Absolute values are hard to give because the resistance of these structures depends on their water content. At the higher end of the scale, one can say that at 125°C at water saturation, that is, the conditions within an operating autoclave, all fungal cells are killed; dry sterilization needs temperatures up to 160°C. At the lower end of the range, fungi can be stored in liquid nitrogen for years, as is routinely the case in culture collections.

In the case of physiological temperatures, one usually cannot give very precise figures for a particular fungus for the range in which growth and development are possible. Indeed, it is more interesting to consider those temperatures at which the fungus is most competitive. Two cases will be considered here: those fungi with maximal activities at unusually high temperatures and those at unusually low temperatures.

8.2 Psychrophilic and cryophilic fungi

There is a mistaken, but general, opinion that fungi prefer warm temperatures. This view has gained credence through the use of high temperatures (> 25°C) in laboratory investigations and the choice of what are essentially thermophilic fungi, for example, *N. crassa*, as the experimental organism. In contrast, some fungal species have their maximal activity around 0°C. Here we consider two types of fungi that flourish at this sort of temperature, predacious fungi, which feed on nematodes, and lichens.

Predacious fungi, like other soil fungi, need a good water supply both for their own activity and that of their prey. In the temperate, subarctic and arctic regions, this is only likely in the winter or rainy season, when many soil organisms have their maximal activity. This has been shown to be the case with nematode-destroying fungi; for instance, *Verticillium suchlasporium*, isolated from soil in southern Sweden, is capable of growth and infection of nematode eggs at 6°C, other predacious fungi have higher temperature optima. Numerous strains of nematophagous fungi have been described from moss cushions of Antarctica, which were mostly frozen and only thawed for some hours per day during the summer period.

Also in Antarctica, lichens have been reported to show gas exchange even in locations which, even during an antarctic summer, remained frozen for most hours of the day. In part this exchange was due to photosynthetic activity of the photobiont but fungal activity also occurred under these conditions.

8.3 Thermophilic fungi

A good site for fungi capable of growing at elevated temperatures is a bakery. Here is found one of the world's most famous fungi, the red mould, *N. crassa*. It reaches its highest rates of extension at about 28°C and shows growth and conidiation up to 35°C. This growth and development at temperatures often used in the laboratory has made this fungus very attractive experimentally. In similar manner, fungi used for industrial purposes are thermophilic, or mutants or strains selected because they show maximum growth at elevated temperatures, since the use of such fungi reduces (in comparison with a fungus only able to grow at lower temperatures) the amount of coolant required to extract the heat generated by fungal activity in a large fermenter. The strain of *Phanerochaete chrysosporium*, now being used for the study of the mechanism of lignin breakdown (see p. 46), is an example of a strain selected for its thermophilic properties, since it is hoped that the same strain will eventually be used industrially.

Typical thermophilic fungi tolerate even higher temperatures. Some of them have been isolated from compost. During composting, temperatures up to 80°C are reached due to fungal and bacterial activity. Fungi have been shown to be active up to 65°C. A list of fungi isolated from commercial composting plants is given in *Table 8.1*.

Some of these fungi, together with bacteria, have been known for a long time because of their ability to generate high temperatures, particularly where heat is not easily dissipated. This can be the case with improperly dried hay. The fungi (and bacteria present) use the available water in the hay to start growing and bring about breakdown of the hay. In this process the microorganisms generate heat, which remains within the body of the hay because of its very effective insulation properties. As the metabolic activity continues at a high rate, the temperature rises such that it reaches a value leading to the self-ignition of the hay, which, due to its fine structure,

Table 8.1: Thermophilic fungi, capable of growing at 50°C, isolated from self-heating wood chips

Aspergillus fumigatus
Chaetomium thermophile var. *coprophile*
Chaetomium thermophile var. *dissitum*
Humicola grisea var. *thermoidea*
Humicola lanuginosa
Sporotrichum thermophile
Talaromyces emersonii
Talaromyces thermophilus
Thermoascus aurantiacus

has a very low energy of activation. Temperatures below 100°C are sufficient to cause the eventual conflagration. A number of fungi listed in *Table 8.1* are known to be involved in the heating process.

8.4 Temperature change

In nature, fungi do not grow in environments in which the temperature is constant, but are subjected to temperature oscillations. These occur during the day but in a bulky structure, like a tree trunk or the soil, the oscillations over a year are more significant. Temperature changes are likely to reach different parts of the colony at different times, bringing about differential rates of extension. This results in some hyphae with negligible rates of extension and, in an analogous manner to cessation of growth with light, sporulation is induced. In the laboratory, it has been shown that a temperature difference of as little as 1°C is sufficient to trigger sporulation; indeed, the same fungi can be induced to form rings in a regimen oscillating about this temperature difference. Fruiting of *Marasmius oreades*, which forms fairy rings, will occur several times during the growing season, the cause thought to be temperature alterations between cool (rainy) and hot (sunny) periods.

Chapter 9

Competition: living with neighbours

9.1 Introduction

In the previous chapters, the reactions of an individual mycelium to environmental factors are described. The text can be justified, as long as axenic fungal cultures are being considered, but it is obviously an oversimplification with respect to a mycelium spreading into its natural substrate to face the competition of all types and numbers of other organisms. The growing mycelium must cope with the challenge of competition or not succeed.

9.2 Competition and the ecological niche

Competition or the act of rivalry between competitors or rival organisms are necessarily global terms, since they comprise all those organisms that are fighting for space, nutrients, water and oxygen and the ability to express their developmental possibilities. The permanent requirement to withstand competition or die forces a fungus either to minimize that competition or to enhance its own competitive ability. Thus, every fungus, like any other organism, has found ways to exploit its substrate in the face of competition, that is, the kind of lifestyle that a fungus has adopted is the result of adaptive processes to increase the probability of survival.

Once the lifestyle and developmental pattern have become established as a result of adaptation, the organism will retain the necessary characteristics, perhaps becoming even more specialized as a result of continuing competition. The lifestyle of the fungus, together with the necessary adaptive features and the resources used by the fungus as it develops, is called the 'ecological niche'. This term does not mean just the space occupied in a particular substrate but characterizes the manner in which an organism has adapted to competition.

9.3 Types of competition

For competition to be properly comprehended, the various aspects need to be considered separately. If we consider competition for nutrients, we must consider it in relation to what happens to a hypha as it extends through the substrate, taking in those nutrients that are already accessible or after they have been hydrolytically degraded by extrahyphal enzymes. When extrahyphal enzymes are involved, there is a difference in time and space between the process of hydrolysis and the transport of the products of hydrolysis into the cytoplasm. The difference gives the opportunity to other microorganisms to have access to these products (*Figure 9.1*). These microorganisms also use these nutrients rapidly and are, in consequence, always found in the immediate neighbourhood of growing mycelia.

Competition at the level of individual hyphae is mirrored at the level of mycelia, when the same or ecologically similar species invade the same substrate. One can see the

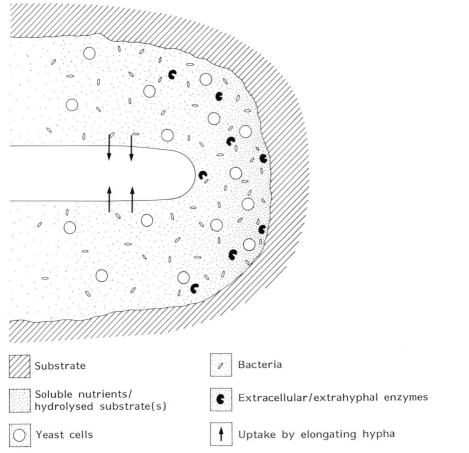

▨ Substrate		₀ Bacteria	
▦ Soluble nutrients/ hydrolysed substrate(s)		☾ Extracellular/extrahyphal enzymes	
◯ Yeast cells		↑ Uptake by elongating hypha	

Figure 9.1:   Diagram showing competition by bacteria and yeast cells around the extending hyphal tip.

result of such competition in wood colonized by *Coriolus versicolor*, the individual colonies of which form black zone lines, where hyphae of the different mycelia have interacted with each other to prevent further movement of each mycelial margin in the region of contact (*Figure 9.2*). The zone lines are therefore barriers to competing mycelia, which essentially seal off the occupied space from greedy neighbours.

That space which is occupied by a fungus provides the nutrients, water and oxygen. In addition, the space is the launching pad for reproduction from the surface through the development of fruiting structures. Provided sufficient nutrients are captured, the surface area regulates the number of fruiting structures and thus the success of reproduction.

Where sexual (teleomorphic) fruiting organs are formed by dikaryotization, the acquisition of other, compatible (see *Table 11.1*, p. 108), nuclei, is a prerequisite. These nuclei can be obtained from spores or other colonies. Spores transported by air or animal vectors can reach a mycelium at the surface only, hence the surface area also regulates the number of (compatible) nuclei obtained. Thus success in gaining a larger part of the substrate increases the chances of the formation of fruiting bodies and an increased nuclear variability, which in turn leads to better success in reproduction and of the ensuing offspring.

Figure 9.2:   Zone lines (and fruiting organs) produced in wood by mycelia of strains of *Coriolus versicolor*. See text for further details. Photograph by A.D.M. Rayner.

9.4 The competitors

In the frame of heteroecology, most organisms in the vicinity of the mycelium, especially those capable of absorbing nutrients made available by the fungus, will be competitors. There will not only be competition from other organisms but also between mycelia of the same species, and even between different parts of the same mycelium or between closely adjacent hyphae. For instance, those producing propagules, perhaps as a result of being stimulated by light, might consume reserves and act as a sink for nutrients entering the mycelium and in these ways affect those hyphae growing trophically.

We have highlighted in the above paragraph the complexity of competition, indicating the hierarchy of situations from that between dissimilar species to interhyphal competition, but having said that, nearly every substrate accessible to fungi can support more than one species. This is obvious with soil but also is evident for timber, where one frequently finds an attack by more than one fungus. Here the case can be illustrated even more strikingly in the infection of *Capsella bursa-pastoris* by *Albugo candida* and *Perenospora parasitica* at the same time and on the same organs (*Figure 9.3*). One might anticipate that competition must still be occurring, with the two species having evolved strategies to minimize its effects. Possible ways in which competition can be minimized are indicated in the following section.

9.5 Escaping competition

Every species of fungus has its own strategy to withstand competition to live and reproduce. These individual strategies or adaptations to ecological niches can be divided into three broad categories.

Figure 9.3:   *Capsella bursa-pastoris* infected by *Albugo candida* and *Peronospora parasitica*.

9.5.1 Combative strategy (C-strategy)

The combative strategy enables the fungus to defend those resources that have already been captured or to attack those competitors occupying a resource that is capable of capture. The classic case is *Oudemansiella mucida*, a basidiomycete inhabiting standing but dead beech (*Fagus sylvatica*) trunks. This fungus is common throughout Europe. It is well known that a beech trunk invaded by *O. mucida* never hosts any other wood decay fungus. This is now known to be due to the production of the antifungal compound, mucidin, which is active against all types of fungi and is used to cure dermatomycoses. There are other examples of ecologically combative fungi which produce antifungal compounds (e.g. *Trichoderma viride,* which produces viridin).

9.5.2 Ruderal strategy (R-strategy)

Ruderal strategy allows the exploitation of substrates as yet unoccupied, or only partly occupied. Such situations are created by the production of newly formed substrates; fresh dung or deciduous leaves or weakened living organisms, which are all susceptible to exploitation. The fungi which invade these substrates are more opportunistic than their competitors. Many zygomycetes use this strategy; they do not attempt to attack any potentially resistant substrates but attack substrates where there is a sufficiency of low molecular weight compounds that can be readily absorbed. The fungi we are considering grow on dung balls and rotten fruit and also on prepared foodstuffs such as bread. Given sufficient water, these fungi can extend at a rate of up to 6 cm a day. So they grow faster than possible competitors and are therefore the first to reach uncolonized substrate (*Figure 9.4*). Since these fungi only

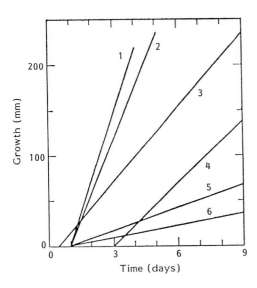

Figure 9.4:   Diagram showing the growth rates of fungi obtained by growing them on nutrient agar in tubes in order to obtain measurements over relatively long time periods. (1) *Neurospora crassa,* sugar fungus; (2) *Rhizopus oryzae,* sugar fungus; (3) *R. nigricans,* coprophilous fungus; (4) *Sordaria fimicola,* coprophilous fungus; (5) *Fusarium oxysporum lycopersici,* saprotrophic fungus or plant parasite; (6) *Monilinia fructicola,* plant parasite. Reproduced from ref. 1 with permission from John Wiley & Sons.

take up readily consumable compounds, being unable to break down insoluble substrates, they are called 'sugar fungi'.

On the other hand the exploitation of very specialized substrates or unusual nutrient sources in a patchy environment is a ruderal strategy equivalent to the one just described. The particular strategy being considered here is present amongst the coprophilous fungi. Amongst these fungi *Piptocephalis,* known as a typical zygomycete from dung balls, is actually a parasite of mycelia of other zygomycetes like *Mucor* or *Rhizopus.* Nematodes and other animals in dung are used as nutrient sources by the predacious fungi *Arthrobotrys* or *Harposporium;* the eggs of nematodes are attacked by *Rhopalomyces.*

Other well known examples of fungi using exotic substrates are *Hymenoscyphus* spp., which grow especially on the veins in the dead leaves of beech or oak, and the genus *Rutstroemia,* which grows on the cupules of fagaceous fruit. This list can be continued with those fungi which are found without exception on former fire sites, such as *Pholiota carbonaria* or various ascomycetes. We still do not know if these fungi use the substrates previously altered by fire or whether the fungi are such weak competitors that they need to have the environment pre-sterilized before they invade it; the consensus tends to favour the latter possibility.

This specialization in the use of substrates has an equivalence in the sequence of fungal species that can be found at different stages in the breakdown of a complex substrate. What one sees on such substrates is a succession of different fungal species, each species using a different component of the system as it becomes available as a result of the degradative activity of the preceding fungi (*Table 9.1*).

9.5.3 Stress-tolerant strategy (S-strategy)

The stress-tolerant strategy relates to those circumstances where the environment is too harsh for most of the possible competition. The fungi concerned are those which

Table 9.1:   The temporal dominance of fungi on dung in relation to the substrates that they utilize

State of degradation	Ecotype of fungi	Names of typical fungal genera
Fresh dung	Sugar fungi Mycophilous fungi	*Mucor, Penicillium, Pilobolus* *Piptocephalis, Syncephalastrum*

Soft structures attacked	Coprophilous ascomycetes
	Predacious fungi (on protozoa,
	  nematodes, etc.)
		Podospora, Saccobolus, Sordaria
		Arthrobotrys, Doratomyces, Harposporium,
		  Rhopalomyces

Woody structures attacked, marked weight loss	'Dung'-basidiomycetes	*Coprinus, Panaeolus*

Formation of humates; dung
  worked into the soil
	Fungi on soils with high
	  organic content
		Anthurus, Coprinus, Cyathus, Phallus

are adapted to low or high temperatures, lack of oxygen or reduced water availability. The reader is referred to the relevant chapters to learn how the particular species are adapted to the conditions just listed.

9.6 Strategies, selection and evolution

The successful development of one of the above strategies can fix a fungus in its ecological niche and allow further adaptation. In terms of the generations that follow, the progeny move further along the direction taken by those ancestors that evolved the initial strategy. The adaptation which takes place is achieved through competition such that only the fittest progeny, that is, those optimally adapted at the time, survive and reproduce. These progeny are thus selected.

This is not the complete story. The genetic flexibility (variability) found in fungi, like all organisms, continuously supplies the population, which is being subject to selection, with new genotypes. When exposed to changed environmental circumstances, some of the new genotypes will survive but others, and these will be deemed unfit, will not. In this way a line of progeny with different environmental capabilities will arise. This continuous process of selection and initiation of new progeny lines results in the creation of new forms, even new species. This is evolution. The reproductive systems and nuclear behaviour that underpin maintenance and transmission of the genetic diversity on which evolution may act is considered in the final part of this book.

Reference

1.   Cochrane, V.W. (1958) *Physiology of Fungi*. John Wiley & Sons, Chichester.

Chapter 10
From trophophase to idiophase

10.1 Introduction

Reproduction involves a change in morphology, whether it be an individual hypha or part of the mycelium. In both cases, there is a change from trophic growth. We can speak of the change as being one of differentiation, that is, the hypha or mycelium is taking on a different function or morphology than hitherto. Both types of change and also any change in vegetative structure of the mycelium, such as the formation of a rhizomorph, sclerotium or fruit body means a change in cell chemistry. When trophic growth ceases due to exhaustion of a nutrient or nutrients, some of the pathways of primary metabolism will either be severely reduced or cease to operate altogether, as in the normal vegetatively growing mycelium; metabolism can be considered to have become unbalanced. It is under these conditions that new pathways of metabolism, not concerned with producing the compounds necessary for growth, namely secondary metabolism, become operative. Because there is this change in cell chemistry, secondary metabolism must also be considered a process of differentiation. The production of secondary metabolites is of great interest, both ecologically and industrially, but understanding how secondary metabolism is switched on can provide clues generally about how differentiation occurs in fungi. The developmental state which the mycelium enters after trophophase, and in which secondary metabolism or reproduction commences, has been termed the idiophase.

10.2 Secondary metabolism

When nutrients, including oxygen, are not limiting growth, a mycelium will grow exponentially and be essentially of uniform composition. When a nutrient or nutrients are exhausted or reduced to a very low level, the growth rate, in terms of production of biomass, is greatly reduced, and indeed may become zero (*Figure 10.1*). It is under these conditions that secondary metabolites are produced (*Figure 10.2*; *Table 10.1*). These compounds cannot be described simply in terms of structure because they are many and varied, nor in terms of function because for the great majority we have no

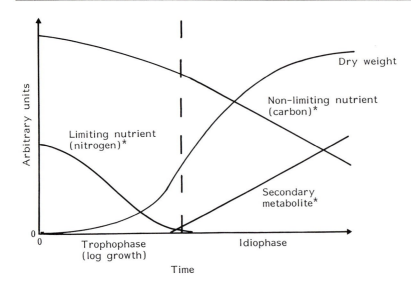

Figure 10.1:   Growth of a filamentous fungus in stirred liquid culture showing the principal changes in the medium during trophophase (balanced or exponential growth) and idiophase. The increase in dry weight during idiophase does not mean an equivalent increase in protoplasm; it is almost certain that increase is due to carbon incorporated into storage material. * Concentration in the medium.

understanding of the role they play in the life of the fungus. The best definition is essentially negative, that is, that a secondary metabolite is not essential for the vegetative growth of the producing organism in pure culture.

It has been customary, because of studies on secondary metabolism started using batch cultures, to consider secondary metabolite production as being a growth-related phenomenon, but we need to remember that, in batch culture, reduction of growth rate is associated with nutrient exhaustion. Furthermore, the characteristics of the external medium do not necessarily give a proper picture of the internal milieu of the fungus that is responsible for the change to secondary metabolism. For secondary metabolism to occur, there must be some process altering the expression of the genome. So it is not surprising to find that secondary metabolite production is enhanced when particular nutritional controls are lifted, examples being repression by glucose, nitrogen and phosphate. In certain instances, however, secondary metabolite production may occur as a result of induction, an example being the stimulation of cephalosporin C production by *Cephalosporium acremonium* by methionine. In keeping with the viewpoint just expressed, it is possible in batch culture to change a mycelium, which is non-growing, to the growing state, albeit with a very low growth rate, by changing the external medium in a defined manner and have a concurrent growth-associated production of a secondary metabolite. Of course, when the mycelium enters the secondary metabolic state it is the production of new

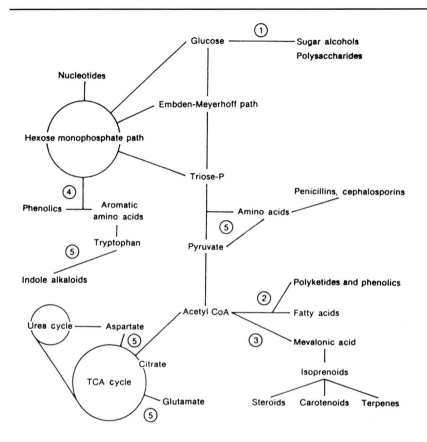

Figure 10.2: The interrelationships of the metabolic pathways in primary and secondary metabolism. The principal pathways of secondary metabolism are numbered as follows: 1, glucose-derived metabolites; 2, acetate–malonate pathway; 3, mevalonic acid pathway; 4, shikimic acid pathway; 5, amino acid-derived pathways. Reproduced from Griffith D.H., *Fungal Physiology.* Copyright © 1981 John Wiley & Sons Inc. Reprinted by permission of John Wiley & Sons Inc.

Table 10.1:   Some important secondary metabolites and the fungi that produce them.

Substance(s)	Fungus
Aflatoxin	*Aspergillus flavus*
Carotinoids	*Phycomyces blakesleeanus*
Cephalosporin C	*Cephalosporium acremonium*
Ergotamines	*Claviceps purpurea, C. paspali*
Ligninase	*Phanerochaete chrysosporium*
Melanin	*Aureobasidium pullulans*
Mucidin	*Oudemansiella mucida*
Penicillin	*Penicillium notatum*

enzyme(s) that is the significant event. A most important example is the production of ligninase, which only takes place in the mycelium of *Phanerochaete chrysosporium* when it changes to the secondary metabolic state. Production of the enzyme is repressed by exogenous carbon and nitrogen.

Secondary metabolites are assumed to play an important role in nature. Knowledge based on experiments or observations is scarce and hence generalizations are difficult. One example is alkaloid production by the ergot fungus *Claviceps purpurea*. The alkaloids produced and stored in the mycelium and the sclerotia may be metabolic reserves but they are also deterrents to animals feeding on the higher plant host. This is similar to the effect of the endophyte in the grass *Lolium temulentum* described previously (see p. 36). Melanin pigments, if deposited in the hyphal wall act as a protection against light, as is the case for *Aureobasidium pullulans,* or in the spores of coprophilous fungi which are ejected on to higher plant shoots (see p. 87), where they are exposed to potentially damaging high light intensities. In long-lived spores or sclerotia, melanins protect the protoplasm from microbial attack and probably help to reduce water loss.

Of course, fungal secondary metabolites are of great importance to man. Some of the important ones are listed in *Table 10.1*. It is this ability to produce such secondary metabolites that contributes to the industrial importance of fungi.

10.3 Morphological changes

Microscopically, the change from trophophase to idiophase is characterized by the disappearance of the Spitzenkörper, a key component of the extension process. Once it has disappeared, it cannot be rebuilt or re-established by that hyphal tip; further extension is only possible by forming a new tip and new branches. The start of reproduction is bound up with the cessation of hyphal extension. This occurs when hyphae reach the edge or the surface of their substrate. If we consider natural conditions, the hyphae will be brought to a halt by reaching the surface of the soil or wood, by a barrier like a stone in the soil or bark on wood or by coming into contact with another colony of the same species. Cessation of hyphal extension as a consequence of reaching the boundary of the substrate may be reinforced by changes in other factors, which may alter at the same locale. This will certainly be true of hyphae reaching the soil–air interface, where, as has already been pointed out (see p. 81), light–dark changes as well as temperature and humidity changes might all act as signals to initiate the development of reproductive structures.

The developmental switch will be accompanied, if occurring on those substrates that have been fully colonized and therefore probably lacking key nutrients, by a metabolic switch, in which endogenous reserves are called upon to support the morphological changes which are about to take place. The reserves or their breakdown products may be brought some distance in the mycelium by translocation (see p. 24). The utilization of reserves leads to an enhanced oxygen consumption, which has often been found at the start of reproduction.

It is with these changes of metabolism that morphological changes take place in most fungi. Typically, aerial hyphae are produced, which are dependent on translocation to supply the necessary nutrients for extension. These aerial hyphae characteristically have water droplets on their surface, presumably as a result of the hydrostatic pressure generated inside. They eventually differentiate into spore-producing organs.

When a large reproductive structure such as a mushroom is pushed into the air, the mycelium producing it is likely to be within a substrate of very large volume. This means that, although there might be local exhaustion of key nutrients, it is possible for the deficiency to be made up by translocation, not only of reserve products as above, but absorbed nutrients from distant parts of the substrate where the same nutrients are just being exploited. With mushrooms, their development is a two-stage process (*Figure 10.3*). First a mycelial mass is produced by non-trophic extension of hyphae in a coordinated manner. The mycelial mass then expands by an increase in hyphal volume and accompanied by a uniform expansion of the hyphal walls.

(a)   (b)

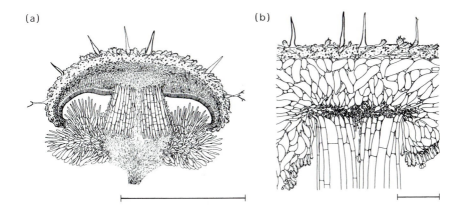

Figure 10.3: Non-trophic growth in fungi. The increase in the size of the fruit body of *Mycena* sp. by an increase in hyphal volume. (**a**) A section through the button stage to show the full preparation of the structure before it expands prior to shedding of spores. (**b**) Central part of the fruit body now fully expanded. Bar represents 1 mm. Reproduced from ref. 1 with permission from The Orion Publishing Group Ltd.

References

1. Corner, E.J.H. (1964) *The Life of Plants.* Weidenfeld & Nicholson, London.
2. Griffin, D.H. (1981) *Fungal Physiology.* John Wiley & Sons, Chichester.

Chapter 11
Nuclear relationships

11.1 Introduction

In fungi, both sexual and vegetative propagation are equally common. In most classes of the lower fungi, oomycetes, chytridiomycetes and (in part) the zygomycetes, these two processes are part of the generation cycles which are comparable to those known from green plant groups.

Among the above fungi, there are typical diplontic organisms, that is those which form a diploid mycelium during its vegetative phase, represented by the oomycetes. There are also typical haplonts with a haploid mycelium and nuclear fusion and meiosis, one following the other very closely. However, even in these groups two typical fungal features are found: male and female structures are formed on all mycelia. Thus there is no differentiation into male and female mycelia. Instead there is differentiation into two (or more) types of mycelia, which can only fertilize mycelia of the opposite or any other type. These types are called 'mating types' to avoid the terms 'male', 'female' or 'sexes'. They are assigned the notation '+/–' or 'A/B', which is interpreted that only '+' can fertilize '–' and similarly for 'A' and 'B'. To describe the capability for mutual fertilization one uses the term 'compatibility'. Compatible mycelia are of the opposite mating type; incompatible mycelia are of the same mating type. For definition of these and other terms see *Table 11.1*.

The other typical fungal feature in the life cycles in the groups above is the tendency to enhance vegetative reproduction and to reduce and exclude sexuality. As an obvious result, many fungi of these groups, and indeed also in the ascomycetes and basidiomycetes, do not form any sexual propagules (meiospores) but form numerous asexual propagules (mitospores). This tendency is a result of the fungal lifestyle. *Table 11.2* lists the differences between sexual and asexual reproduction with reference to the ecological relevance of each feature considered. It can be seen that selection has worked on most fungi in the direction of asexual reproduction. Nevertheless, in the long term, asexual reproduction by itself would be detrimental to evolution, therefore fungi combine the advantages of asexual reproduction and sexual propagation at one and the same time, namely by dikaryotization and parasexuality.

Table 11.1:   Terms describing the mating and nuclear relationships in fungi

Term	Relationship
Monokaryon (adj. *monokaryotic*)	Mycelium containing nuclei of only one mating type
Dikaryon (adj. *dikaryotic*)	Mycelium containing nuclei of both or two mating types
Homokaryon (adj. *homokaryotic*)	Mycelium containing nuclei with only one allele with regard to a distinct gene which exists in several alleles
Heterokaryon (adj. *heterokaryotic*)	Mycelium contains nuclei with different alleles of a distinct gene
Homothallic	Self-fertile, no second mating type required for sexual reproduction
Heterothallic	The species needs two different mating types for sexual reproduction
Pseudohomothallic	From the very beginning (i.e. from the formation of the spore) there are nuclei of both mating types present and hence sexual reproduction may occur at any favourable time or developmental state (e.g. four-spored ascomycetes or two-spored basidiomycetes)
Mating type	A group of nuclei/strains which cannot react sexually with one another but need the opposite (+/− system) or an unlike ($A_n B_n$ systems) nucleus for reproduction
Compatibility (adj. *compatible*)	Sexual reproduction is able to occur between mycelia of different mating type
Incompatibility (adj. *incompatible*)	No mating takes place between incompatible colonies because: with *homogenic incompatibility* both nuclei involved are of the same mating type with *heterogenic incompatibility* possible mating is prevented by other (so-called 'incompatibility') genes
Anamorph	The term for the form of reproduction, in this case vegetative (giving the conidial or imperfect state), and also the taxonomic name of the species with this form of reproduction
Teleomorph	The term for the form of reproduction, in this case sexual, and also the taxonomic state of the species with this form of reproduction
Holomorph	Anamorph + teleomorph, which according to the nomenclatural rules takes the taxonomic name of the teleomorph
Barrage	The form of mutual killing when mycelia meet which are incompatible. Pigmented colonies show a white line where the meeting occurs; in wood attacked by mycelia of white rot fungi, dark lines appear

11.2 Dikaryotization

In the ascomycetes and basidiomycetes, only mycelia containing two compatible nuclei are capable of sexual reproduction by forming asci or basidia (see Life Cycles 3, 4 and 5). In ascomycetes, this dikaryotization occurs in the fruiting organ initials, which take up compatible nuclei by various means (*Figure 11.1*). The nuclei which arrive either move through the vegetative mycelium and through the septal pores (*Figure 11.2*) or are incorporated together with the original nucleus into dikaryotic 'ascogenous' hyphae, which maintain their dikaryotic state. These hyphae grow between the non-ascogenous, monokaryotic hyphae and later on form the asci. One consequence of this mode is that the entire mycelium or individual fruiting organ

Table 11.2:   The possible successful outcomes from reproductive activity and the extent to which they are achieved by sexual or asexual propagation

Target	Sexually achieved	Asexually achieved
Starting a new colony	+	+
Finding a new substrate	+	+
Germinating at the proper time	+	+
Resting periods possible	+	+
Genetic recombination	+	(−)
Multiplication of successful genotypes	−	+
Providing additional nuclei for dikaryotization or heterokaryosis	+	+
Starting reproduction very quickly	−	+
Enhance quantity of reproductive units	(−)	+
Minimize consumption of resources	−	+

initial can be dikaryotized several times by nuclei of different origin. The resulting ascogenous hyphae thus differ widely in their genetic pool; these hyphae are thus said to be heterokaryotic.

In ascomycetes, there is a tendency to extend the dikaryotic phase; as shown in *Figure 11.3*, in four-spored ascomycetes like *Podospora anserina* or *Neurospora tetrasperma* nuclei of two compatible types are always incorporated into one spore. The germinating hypha and subsequent mycelium is thus dikaryotic from its very inception, capable of starting reproduction without the need to delay for the arrival of compatible nuclei.

1 Conidium, attached to trichogyne   5 Spermatium (microconidium)

2 Conidium delivering nucleus   ○ Nuclear type present in the mycelium
   (nuclei) to mycelium
                                          ● Dikaryotizing nuclei (compatible nuclei)
3 Hyphal contact
                                          → Nuclear migration
4 Ascospore

Figure 11.1:   Different methods of gaining compatible nuclei for dikaryotization in ascomycetes.

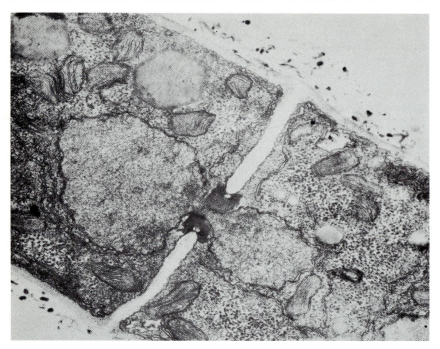

Figure 11.2: Electron micrograph of nuclei migrating through a septal pore in a hypha of *Neurospora crassa.* Photograph by M. Bourne by kind permission of Sir John Burnett.

In basidiomycetes, the dikaryotic phase is considerably extended; germ tubes of basidiospores or conidia fuse as soon as they come into contact, exchanging their nuclei. Since often a spore germinates close to others, these hyphal fusions and dikaryotizations are multiplied, leading to different pairs of nuclei. The germination behaviour of basidiomycetes favours dikaryotization. In *Leccinum* and other genera a mutual stimulation of germination has been found, which can be interpreted as a means of increasing hyphal contacts and thus dikaryotization. The deposition of the major proportion of the spores that are produced in the immediate neighbourhood of the parent mycelium, as is often the case with mushrooms (*Figure 11.4*), cannot be understood without considering the probable genetic consequences.

Although the germination phase may be completed, there is a continuing incorporation of nuclei. Even dikaryotic hyphae continue to incorporate further nuclei. This is achieved by hyphal contact and subsequent fusion. In addition, basidiospores or conidia (mono- or dikaryotic), which settle close to a mycelium of the same species, come into contact with the growing hyphae, which incorporate the nuclei of the propagules. Thus dikaryotization, or the absorption of more nuclei, occurs during the entire lifespan of the mycelium (*Figure 11.5*).

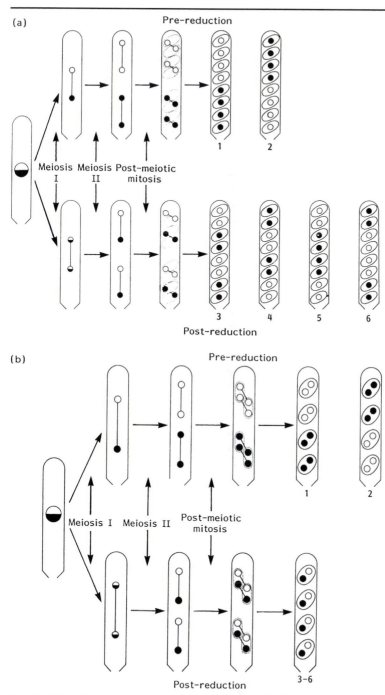

Figure 11.3: Diagram showing the distribution of different genomic markers (o and ●) by pre- and post-meiotic reduction in four- and eight-spored ascomycetes. See ref. 1 for further details.

Figure 11.4: Photograph of fruit bodies of *Armillaria* sp. to show how the spores from one cap can fall on to another cap (arrows), resulting in very limited dispersal of the spores. Photograph by H. Michaelis.

On the other hand, in basidiomycetes there are also mechanisms to maintain the dikaryotic state; this is achieved through the formation of clamp connections (see Life Cycle 5). These are associated with the formation of a septum and the formation of two hyphal compartments where only one existed. The clamp connection ensures that both of the new compartments are in the same dikaryotic state as the original compartment. Clamp connections can be considered to be analogous to hooks of ascomycetes (see Life Cycle 3). Sometimes there are 'whorl clamp connections', multiple clamps around one septum; probably they are related to different nuclear types in one hypha or mycelium.

What are the advantages to the mycelium of this behaviour of absorbing nuclei, even after dikaryotization has been achieved, but remaining in the vegetative state without forming diploid nuclei or switching to sexual reproduction? The answer may be two-fold and must be regarded as speculative.

The dikaryotic state means the existence of independent nuclei in the one mycelium. This independence comprises nuclear movement and division, the presence of different alleles and different exchangeability against newly entering nuclei. Thus a pair of nuclei never form a relationship, one can be removed by differential movement along the hypha, it can decline relatively by a lower rate of division, which itself may be caused by alleles which are unfavourable under given conditions. This means that in a given mycelium, not only may different pairs be present from its inception, but also there is a reshuffling of nuclei throughout the lifetime of the mycelium. Since the different nuclei are obviously not identical, this means a continuous reshuffling of the genetic pool and thus continuous adjustment to a (changing) or patchy environment.

The dikaryotic state, once established, also allows sexual reproduction to start at any favourable time. The main events of the life cycle like meiosis, nuclear conjugation,

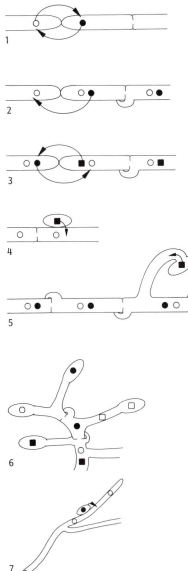

1 Hyphal contact
2 Contact of mono- and dikaryon
3 Contact of two dikaryons
4 Conidium or meiospore
5 Homing of spores
6 Fusion of germ tubes
7 Flexuous hypha of a rust fungus
   with pycnospore (microconidium)
→ Migration of nuclei

Figure 11.5: 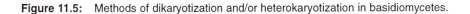 Methods of dikaryotization and/or heterokaryotization in basidiomycetes.

sexual reproduction, and so on, are not arranged in a strict temporal sequence, as is the case in plants or animals. This can be considered an advantage for a fungus growing inside its substrate where there are no temporal cues like light (day or year) or ambient humidity. There is only temperature as an external cue but then only as a delayed and dampened signal. The ripening of gametes and their transfer at a set time is not possible. With dikaryotization, however, fungi (at least basidiomycetes) are capable of starting sexual reproduction at any time that conditions become favourable. Consequently, fruiting organs of basidiomycetes and ascomycetes are rarely produced at a particular time during the year but can, provided the environmental conditions are favourable, be produced at any time. This is exemplified by coprophilous fungi, which grow and propagate when a dung ball is available, independent of season. The fruit bodies of many wood-decay fungi are found throughout the year. Mushroom hunters know the phenomenon that when fruit bodies are prevented from being produced by dry periods in summer and autumn, they may be produced in the following April or May.

11.3 Parasexuality

Parasexuality is a portmanteau term covering all known types of nuclear behaviour in deuteromycetes, fungi without (known) sexuality. It covers several quite independent forms of nuclear or genetic reshuffle.

The first component is the input of nuclei into a mycelium. This begins with the incorporation of more than one nucleus into one conidium, which has been shown in the genera *Penicillium* and *Aspergillus*. The conidia contain up to 12 nuclei, which normally do not belong to the same type. The possibilities of nuclear input into a hypha are enhanced further with those spores that have more than one cell. As with the germination of several closely adjacent conidia, hyphal fusion leads to one single mycelium, which contains all the nuclear types available. In *Arthrobotrys* it has been observed that even conidia still fixed to the conidiophore form germ tubes, which fuse locally (*Figure 11.6*) and consequently contain all the nuclei from the conidial cells. It can be assumed that such nuclear input continues during the lifetime of the mycelium, certainly by a conidium landing on or near a mycelium, which acts as a 'sink' for nuclei.

As indicated above, this input is counterbalanced by a steady loss of nuclei, mainly due to different rates of division, which leads to a preponderance of one type over others. This is often visible as 'sectoring' in culture of fungi (*Figure 11.7*). The sectors mark loss of nuclear types. The resulting mycelium is different and grows more vigorously than before it expressed the phenotypes found in the dominated sector. The repeated occurrence of sectoring in some colonies indicates an unstable combination of nuclear types. The relative number of types can be regulated, as has been shown in *Penicillium*, where the percentage of one particular type of nucleus can be altered by an alteration of the nutrients in the medium.

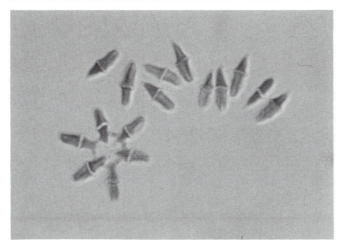

Figure 11.6: Fusion of germ tubes in *Arthrobotrys oligospora*. Photograph by A. Kauflold.

In addition to these gains and losses of nuclei and hence of genetic variability, something termed 'mitotic recombination' has been observed in different fungi. This process leads to the exchange of genetic material between different nuclei, similar to the recombination by meiosis and gamete fusion. How this is achieved remains obscure, since in fungi the nuclear membrane does not disappear during mitosis.

Finally, there is a need to take into account spontaneous mutation as a means whereby the genotype of mycelia may be altered. A fungal colony with a radius of 5 mm and a moderately dense mycelium will contain up to 1.6×10^6 nuclei (*Table 11.3*). Natural mutation rates are in the range of one in every 10^6–10^9 nuclei. Thus any colony with a radius larger than 5 mm will be subject to spontaneous mutation.

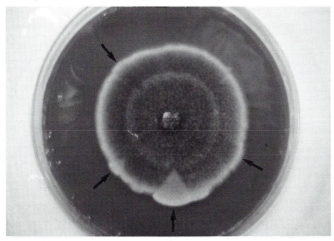

Figure 11.7: Sectoring in a culture of *Cladosporium* sp. growing on nutrient agar. Photograph by M. Kloidt.

Table 11.3:   Calculation of the number of nuclei in a hypothetical colony.

Assumptions	
Number of nuclei per mm hyphal length	20
Distance apart (mm) of the hyphal tips at the surface of a pellet (sphere)	0.05
Therefore	
Hyphal length in 1 mm^3 pellet	300 mm
Number of nuclei in the pellet	6000
Volume of a colony on surface of nutrient-agar (half-sphere with radius of 5 mm)	261.8 mm^3
Therefore number of nuclei in the colony	1.6×10^6

Reference

1.   Esser, K. and Keunen, R. (1967) *Genetics of Fungi.* Springer, Berlin.

Chapter 12

Spores: dispersive propagules

12.1 Introduction

In fungi, reproduction, or propagation, is invariably connected with the production of spores. These can be defined as small parts of fungal colonies, often produced in or at specialized structures and fully equipped to start a new colony independent of the parent mycelium and probably some distance from it. This definition is independent of the details of their production, whether it is sexual or asexual, and it takes no regard of size, shape or special features. The term even includes those conidia, which, as microconidia, act exclusively, or at least mainly, in dikaryotization and thus may also be assigned as spermatia. It is obvious that such a wide definition demands further characterization of the spores according to their type, form and function.

12.2 Types of spores

Fundamentally, one has to distinguish between spores, the nuclei of which are the product of meiosis, meiospores, like ascospores or basidiospores, and those which are produced by mitosis, mitospores. In particular, we need to remember, as discussed earlier, as four-spored ascomycetes or two-spored basidiomycetes show, meiospores are not always monokaryotic nor are they produced for long-range distribution. In terms of ecological function, however, these exceptions can be ignored.

Among the mitospores, those which are motile, the zoospores or planospores, are restricted to the lower fungi, oomycetes and chytridiomycetes. Motility obviously is an advantage in water, where the spores of these fungi are capable of swimming to new substrates. In the long run, however, this possibility is disadvantageous. The activity is restricted to liquid water and, more hampering, swimming itself is an inefficient form of long-range transport. Since velocities are in the range of a few cm per hour and, since no unicellular organism swims in a straight line and the energy supplies last for only a few hours, the distance achieved is miniscule. Transport over distances longer than about a metre thus demands other mechanisms, which is well demonstrated by the oomycetes.

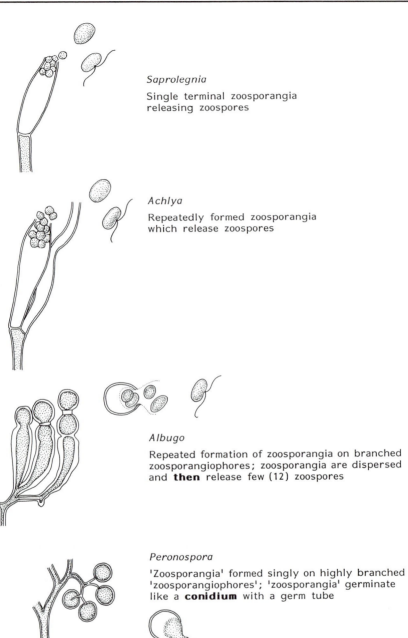

Saprolegnia

Single terminal zoosporangia
releasing zoospores

Achlya

Repeatedly formed zoosporangia
which release zoospores

Albugo

Repeated formation of zoosporangia on branched
zoosporangiophores; zoosporangia are dispersed
and **then** release few (12) zoospores

Peronospora

'Zoosporangia' formed singly on highly branched
'zoosporangiophores'; 'zoosporangia' germinate
like a **conidium** with a germ tube

Figure 12.1:   Diagram showing the steps in the presumed evolution of the conidium in
oomycetes.

12.3 The conidium

The oomycetes have evolved a sophisticated system of zoospore distribution, which in some cases consists of two generations, as is the case in *Achlya* and *Saprolegnia*. The order of Peronosporales, parasitizing living plants, are thought to have migrated in evolutionary time to the land following their hosts. In view of this it can be hypothesized that, at first, these parasites retained their ancestral zoospores but these were formed in zoosporangia, themselves capable of long-range transport. In the genus *Albugo*, the zoosporangia are dispersed and, when they reach a new host, zoospores are liberated, then swim to the stomata and enter the leaf through these natural openings. Later, this dependence on liquid water was lost, a germination tube being formed instead, which can grow to openings (wounds, stomata) and infect the plant without forming zoospores. The ancestral zoosporangia thus have evolved into conidia. An outline of this probable evolutionary development is given in *Figure 12.1*. A similar evolutionary pattern can be seen in the zygomycetes, as shown in *Figure 12.2*. The majority of known fungi produce conidia (they are the most common spore type), which, as shown in *Figure 12.3*, can be divided into two basic types on the basis of their formation: blastic and thallic. Both types can be formed directly at the hyphal apex or, often repeatedly, on small branches called phialides as phialidospores or phialidoconidia.

Numerous variations of both types exist, the simplest being the unicellular spores without any differentiation, as they are formed by many moulds of the genera *Penicillium* or *Aspergillus*. Such spores are formed in huge numbers but distributed randomly, that is, without any targeting, by and through the air or by adhering to the surface of insects and other animals. The high numbers and the distribution by even the smallest movement of air means that these spores are present everywhere. Thus there is a very high chance of moulds being present whenever a suitable substrate is available.

12.4 Dispersal and propagation

Nevertheless in spite of its apparent success, spore production as exemplified by *Penicillium* or *Aspergillus* is only moderately efficient, if one takes into account the huge spore masses required to warrant survival of the species. Therefore, it is not unexpected that means have been developed to enhance success in colonizing new substrates by improving transport, targeting and methods of substrate acquisition. One can start with sculpturing of the spore surface; warts or spines or other appendages enhance friction against the air (or water for aquatic fungi) and thus help to keep the spore drifting in the medium. Appendages also help the spore adhere to a surface of a plant or animal host or other exposed substrates. *Figure 12.4* shows a spore straddled between two protruding cells of the leaf surface.

If the number of cells in a spore is enhanced, still greater variability in structure is possible, as shown by helicosporous and staurosporous conidia (see *Figure 5.1*, p. 71).

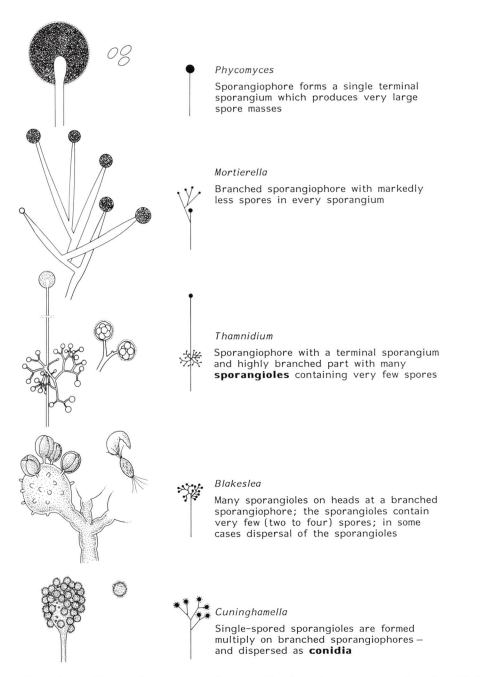

Phycomyces

Sporangiophore forms a single terminal sporangium which produces very large spore masses

Mortierella

Branched sporangiophore with markedly less spores in every sporangium

Thamnidium

Sporangiophore with a terminal sporangium and highly branched part with many **sporangioles** containing very few spores

Blakeslea

Many sporangioles on heads at a branched sporangiophore; the sporangioles contain very few (two to four) spores; in some cases dispersal of the sporangioles

Cuninghamella

Single-spored sporangioles are formed multiply on branched sporangiophores – and dispersed as **conidia**

Figure 12.2:   Diagram showing some of the transition forms between sporangia and conidia in zygomycetes.

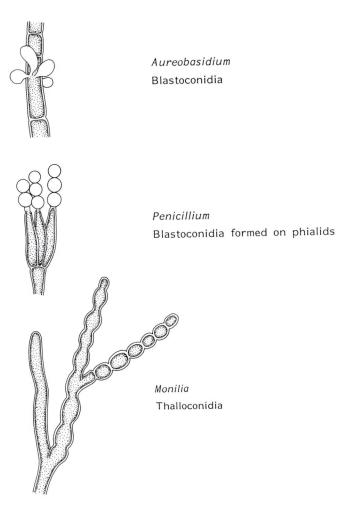

Aureobasidium
Blastoconidia

Penicillium
Blastoconidia formed on phialids

Monilia
Thalloconidia

Figure 12.3:   Blastic, blastic-phialidic and thallic formation of conidia.

While the advantage of the helicosporous form is still uncertain, the shape of stauroconidia helps the spores to float, as has been shown for both freshwater and marine forms (see p. 70). There is also a tendency for such spores to make contact with a solid surface at two points, increasing the amount of anchoring. Staurospores also include the colloquially named tetraradiate spores which are characteristic of freshwater fungi and also found amongst the marine species (see *Figure 5.1d*, p. 71).

Animals can be vectors for fungal spores. This is well known for coprophilous fungi, the propagules of which pass through the intestines of herbivorous mammals and are thus transported from the original (grazing) site to the location of defaecation.

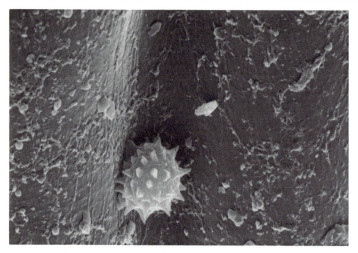

Figure 12.4: A spore hooked between protruding epidermal cells of a leaf. Photograph by U. Stetter.

Arthropods are even more significant. Their outer surface can be loaded with fungal spores when they emerge in spring through fallen leaves, covered with fungal reproductive organs. Insects may also collect spores from sites like rotting flowers, rotting fruit and a wide variety of mouldy substrates. We have already mentioned how yeasts in the nectar of flowers are transmitted to new sources by insects (see p. 79). Fungi enhance their attractiveness to arthropods through secretions, as is the case with pycnidia of rusts and *Claviceps* during its conidial phase. The fragrance of mushrooms attracts flies for oviposition; the young imagoes then take the spores when flying to the next fruit body. The use of insect vectors has meant the development of spores that become either glued by secretions to or, as a result of their shape, entrapped in the cover of hairs or bristles on the insect concerned (*Figure 12.5*).

Another type of spore morphology with a clear function is shown by the genus *Harposporium,* which contains fungi parasitizing small animals. As shown in *Figure 12.5,* the spores found in this genus are hooked or curved and thus straddle between the muscular fibres in the oesophagus or hook into the flesh of these animals. From this point they germinate, the germ tube growing into the body cavities, exploiting the animal until it dies. It would seem that these spores are a successful design. After the total digestion of the parasitized animal, conidiophores grow through the cuticle into the external air and there produce conidia, in most cases not more than several hundred. This number, which is sufficient successfully to maintain the species, contrasts with the tens of thousands (at least) produced by a mould colony.

The success of spores can be enhanced further by attracting host organisms. In *Nemtoctonus* and *Drechmeria coniospora,* the spores possess a sticky knob, which secretes a substance attractive to nematodes, to which these conidia can adhere (see *Figure 3.2*; p. 32). From the point of contact, the germination tube grows into the

Harposporium

Conidium hooks into tissue of oesophagus or intestine of host animal (nematode)

Nematoctonus

Immediately after liberation a short germination tube develops a sticky knob which glues the conidium on to a passing host animal

Cladosporium

Warty or spiny structure allows the spore to become entrapped in the bristles of insects and later may adhere to rough plant surfaces

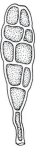

Alternaria

Multicellular spore allows for repeated attempts to germinate (and establish) successfully

Mucor, Pilobolus

Resistant spore walls ensure survival during the passage through the intestine

Figure 12.5:   A selection of spores showing the suggested ecological function of their size and shape.

cavity of the animal, digesting it and exploiting it as long as it is alive. During this time, the animal also transports the fungus to those sites where there is a high probability of there being other nematodes of the same species; hosts become within easy reach of fungal progeny. The nematodes are thus victim and vector in one. The fact that often less than 1000 conidia are formed after digestion is complete gives clear evidence of the efficiency of the reproductive process.

Multicellular spores can also be considered as large spores, since their total length can be as much as 100 µm in the long axis. Increasing size has the immediate consequence that the spores are less readily transported by air, so that other means of dispersal have to be used. Often large spores have warts, spikes or larger protrusions, which allow them to stick to the bristles of insects and other animals. Conidia of the genera *Monoacrosporium* or *Dactyella* sometimes germinate with a sticky knob adhering to the nematode. The animal carries the spore away, at least for some distance, and later the fungus invades and digests the nematode, which provides the nutritional start for the new colony, at the new location.

These larger and multicellular spores often show thick protective walls and are often pigmented. They are formed in large numbers by the mycelia. They act as resting structures, not germinating readily. They are covered by the definition of spores given at the start of this chapter, since they are fully equipped to start a new colony independent of the parent mycelium. This is irrespective of whether or not they are unicellular, like chlamydospores, or larger structures like the spores of *Alternaria*. For the same reason, even sclerotia, those pigmented compact mycelial masses which also act as resting structures and which are referred to previously on page 14, and are visible to the naked eye, also come under the definition of spores.

Chapter 13

Spore germination: starting a new colony

13.1 Introduction

Reproduction cannot be regarded as successful or successfully completed until a new and actively growing colony is established. The last and equally crucial step in the process of reproduction is germination.

13.2 Prerequisites for germination

Before germination starts, some prior conditions must be met. Thus it is obvious that a spore must have developed to a state in which it is capable of germination; it must be mature. In typical conidia, this is immediately after liberation, or even just before, as in *Arthrobotrys oligospora*, where germination can be observed while the conidia are still connected with the conidiophore (see *Figure 11.6*; p. 115). Generally speaking, conidia are ready to germinate immediately after their formation. This ability for immediate germination, however, often does not last for long; conidia of *N. crassa* kept in air lose their germinability after a few weeks.

On the other hand chlamydospores, gemmae, thick-walled telio-, basidio-, asco- or conidiospores are not ready to germinate until they have had a resting period, often months or even years. The thick walls of these spores protect them during the resting period. Such walls also prevent or delay germination and must be weakened before it starts. With spores resting in the soil or inside the remains of the former host, weakening is achieved by weathering or by microbial action. Sclerotia of *Botrytis cinerea* are, in part, degraded by soil microorganisms during winter and early spring. In this way they are activated to start germination in time to infect their host plants when they are starting to sprout.

The thick-walled spores of coprophilous fungi like the ascospores of *Podospora*, *Sordaria* or *Saccobolus,* the conidia of *Alternaria* or *Bispora* and the basidiospores of

Coprinus are, in part, digested during their passage through the intestine. Only after this passage, that is, after partial degradation of the spore wall, can germination start. The projectile mechanisms that many of these fungi have developed (see *Figure 7.6*; p. 87) help to bring the spores to the green parts of the surrounding plants. Here they rest until the herb is eventually eaten by grazing animals and the spores are exposed to digestion. The delay in germination thus ensures that the spores do not germinate early, which could be lethal, and that they germinate and form colonies when they find the optimal condition in the faeces.

But the signal for germination is even more sophisticated than just indicated. The higher temperature inside the animal, mostly mammals, and the drop in temperature when the faeces are deposited are the required cues for germination. This situation is readily mimicked by heating spores to activate germination, which is routinely done in the laboratory with ascospores of *Sordaria,* which do not germinate readily under normal conditions.

Various compounds have been shown to trigger germination, as again is the case for coprophilous fungi. The high ammonium content in rotting dung is important in this respect. Adding ammonium salts (ammonium acetate) to a medium often helps initiate germination in *Podospora anserina,* the spores of which are otherwise difficult to germinate in culture. A similar situation is almost certainly the mutual stimulation of spores. *Leccinum* basidiospores germinate more readily at high density than at low density. This effect is assumed to be the means to ensure dikaryotization and heterokaryotization in developing colonies. A similar reason may lie behind the fact that germination of spores of *Agaricus bisporus* depends on the presence of a specific volatile long-chain fatty acid, isovaleric acid, which is produced by the actively growing mycelium (though the volatile is also produced by other fungi). Just the opposite to what has been reported for *Leccinum* above is found with *B. cinerea;* the rate of conidial germination is inversely correlated to the spore concentration. This has been interpreted as being due to inhibitory substances excreted from the spores. With too many spores, the concentration of this still unknown compound becomes too high and germination is inhibited.

In a similar manner, substances excreted by leaves of *R. nigra* inhibit or prevent germination of spores attached to the surface, which may be a means of regulating the density of fungi in the phylloplane. Chemical inhibition of germination might also account for the phenomenon of soil fungistasis. This term describes the repeatedly found inhibitory effect on fungal spores by and in a soil with established microflora, which makes it virtually impossible to introduce fungi into natural soils by 'sowing' the spores. They do not germinate and are mostly destroyed by all types of soil organisms.

Air pollution is known to affect spore germination. In *Cladosporium,* the germination of conidia on leaves is severely inhibited by dissolved SO_2, with concomitant reduction of numbers of colonies on the leaf. Another fungus that appears to be affected by air pollution is *Rhytisma acerinum*, the tar-spot disease of sycamore (*Acer pseudoplatanus*), which disappears in highly populated or polluted urban areas. This is attributed to the inhibitory effects on germination of SO_2 and other pollutants. The difficulty with this and similar interpretations is that the absence of tar-spot in large cities can also be explained by the considerable disposal by man of fallen leaves, which means the absence of sufficient inoculum in the following spring.

13.3 Environmental factors involved in germination

Spores have to take up water before they start germination. Keeping water out is thus a simple and effective means of preventing or delaying germination and probably is a significant factor in propagules possessing thick walls. On the other hand, although water is indispensable for spore germination, the quantities of liquid water required are quite low. Wetting of leaves in the early morning by dew is sufficient to allow germination of phylloplane and phytopathogenic fungi, even the swimming of the zoospores of *Albugo* and other oomycetes. The conidia of *Aspergillus niger* require no more water than in damp plaster to produce the well known damp stain.

The roles of light and temperature in germination are similar to their roles in hyphal elongation. Usually germination is unaffected by the absence of light or low or high temperatures. However they may sometimes act as a signal for germination. We have pointed out the role of an elevated temperature in triggering germination of the spores of coprophiles. The matter of such signals requires further exploration.

Oxygen is crucial for germination, virtually all fungi need at least microaerobic conditions for the process. During germination there has to be a supply of energy and metabolic precursors for growth. Glycogen, trehalose and lipid are common as reserve substances responsible for that supply. Lipids can constitute between 1 and 13% of the dry weight of a fungal spore and, in electron microscope studies, lipid droplets can be seen to disappear on germination. This evidence for lipids as the reserve material supporting spore germination is supported by other studies showing that the respiratory quotient during germination can be compatible with lipid utilization. It can also be stated that lighter spores are superior to heavier ones in terms of distribution. If we accept this statement, one can see that lipid is the more appropriate storage compound for spores, since it is lighter and contains more energy than polysaccharide or protein. Lipid is, at first sight, more difficult to utilize. This is because, although easily degraded oxidatively to produce energy, the degradation is not accompanied by the production of metabolic precursors. This apparent disadvantage is avoided by the involvement of the glyoxylate cycle or glyoxylate bypass (*Figure 13.1*), which feeds carbon skeletons into the citric acid cycle, from whence they are incorporated into amino acids, soluble carbohydrates and polysaccharides (see p. 63).

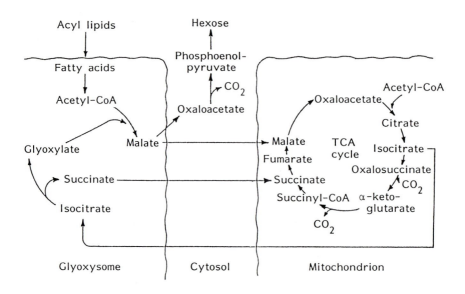

Figure 13.1:   The glyoxylate (isocitrate) cycle.

13.4 The course of germination

In those spores with a water content much reduced compared to that of the parent mycelium, water uptake is required to rehydrate the protoplasm. Of particular importance is the rehydration of membranes, discussed already with respect to the rehydration of lichens (see p. 73; and *Figure 13.2*). Water uptake also activates the utilization of reserve compounds, and if they are insoluble, their soluble breakdown

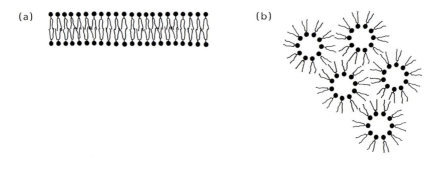

Figure 13.2:   Diagram showing the two conformations formed by amphipolar lipids and water, according to the water content of the lipid. (**a**) More than 20%; (**b**) 15% water; (●) polar (hydrophilic) 'heads'; (≈) apolar (lipophilic) 'tails'. This diagram indicates how dehydration can lead to breakdown of the lipid bilayer of a biological membrane.

products will increase the osmotic activity of the protoplasm, leading to still further water uptake. The enlargement of germinating spores is, in part, due to this water uptake and, in part, due to synthesis of new material, of which new wall material is an important component (see *Figure 13.3* showing the newly formed wall layer in germinating conidia of *B. cinerea*).

As the process of germination proceeds, wall synthesis concentrates at one point, often a pre-formed germination pore, and the germ tube emerges from the spore. The germ tube is organized like a growing hyphal tip, although its metabolism is still based on endogenous reserves. With the elongation of the germ tube, the systems for nutrient acquisition from the external medium commence operation. The switch from endogenous to exogenous nutrient supply is of importance and a pause often occurs at this stage. Growth ceases and specialized organs for nutrient acquisition are produced, such as the appressorium of phytopathogenic fungi (see p. 35). Nematode-destroying fungi may germinate to produce a capture organ, for example, a sticky knob, which adheres to nematodes enabling attack to take place and providing food for the developing mycelium (see *Figure 3.2*; p. 32).

When an external supply of nutrients has been achieved, the germination phase is terminated on the formation of a septum at the base of the germination tube, which is then transformed into an independent hypha (*Figure 13.3*). The hypha continues to extend and starts to branch and the colony develops into the mycelium whose properties have been discussed in the earlier chapters of the book.

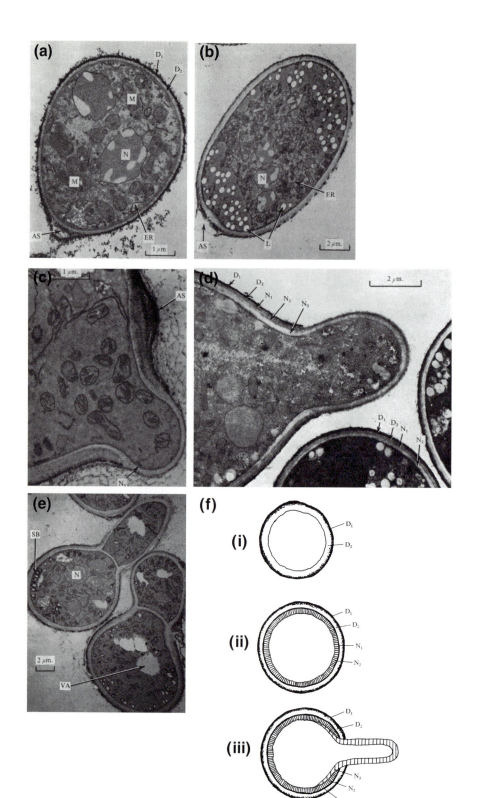

Figure 13.3: Electron micrographs, together with a diagram (all by kind permission of K. Gull and A.P.J. Trinci), showing the course of spore germination in *Botrytis cinerea*. (a) Dormant conidium showing two-layered spore wall. (b) Conidium after 4 h incubation in nutrient medium; a new layer is seen near the abscission scar. (c) Near median section of an incipient germ tube after 10 h in the nutrient medium showing new wall layer. (d) Newly emerged germ tube showing new wall layers, which can also be seen in the other spores. (e) Two fully germinated spores showing the septum at the base of the germ tube and the vacuolization of the spore and germ tube. (f) Schematic representation of the changes in wall structure during germination: (i) dormant conidium; (ii) swollen conidium 4–6 h after inoculation; (iii) conidium with germ tube approximately 10 h after inoculation.

(a), (b), (c) and (e) $KMnO_4$-fixation; (d) glutaraldehyde-fixation. AS, abscission scar D_1, D_2, layers of the spore wall; N_1, N_2, N_3, new wall formed during germination; ER, endoplasmic reticulum; M, mitochondrion; N, nucleus; SB, storage body; VA, vacuole.

Five representative fungal life cycles

These life cycles are represented by two diagrams. (**a**) represents those morphological changes that occur during the life cycle, giving some idea, particularly, of the nature of the reproductive structures. (**b**) is designated a cytological diagram showing nuclear changes, particularly the occurrence of meiosis (R! – reduction division) and nuclear fusion (K! – karyogamy). In these particular diagrams, nuclei may be represented by different shapes, both solid and hollow, to indicate their different genetic makeup. P!, plasmogamy; M!, mitosis.

(a)

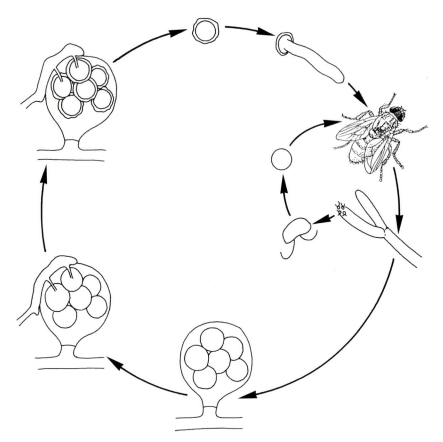

Life Cycle 1: an oomycete water mould, *Saprolegnia*

This (simplified) life cycle is for a truly diploid fungus. The germination of the zygote, which can be considered as a spore, leads to infection of animals like insects and fish, both living and dead. The invading mycelium exploits the animal tissues and extends into the surrounding medium, where it forms terminal zoosporangia filled with zoospores (planospores). After release, the zoospores infect another host either directly or (in some species) after losing their flagella and encysting, and, after some time, forming secondary zoospores. Zoospores are again formed on the new host. Later, antheridia and oogonia are formed (their formation is mutually regulated by hormones) and meiosis takes place within them. After meiosis (R!), mitosis follows and, in the oogonium, several oospheres and, in the antheridium, male gametes are

(b)

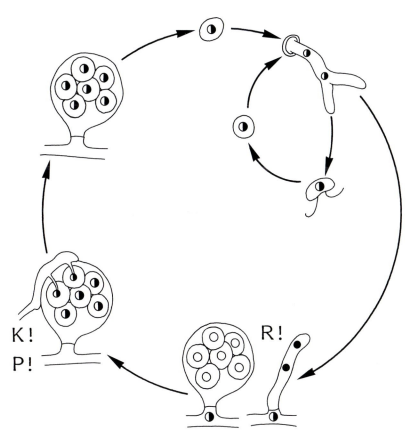

K!
P!

R!

formed. The antheridia then grow into contact with the oogonia and also form fertilization tubes in which male gametes are transported to the oospheres and fertilization takes place. After fertilization, both nuclei fuse (K!) and the zygote formed becomes covered with a thick wall. The zygospore, as it is now called, is relatively long-lasting. Under suitable conditions it germinates to produce a hypha, which infects a new host. As one can see, diploid nuclei are present for most of the life cycle, while haploid nuclei exist only for a short time.

(a)

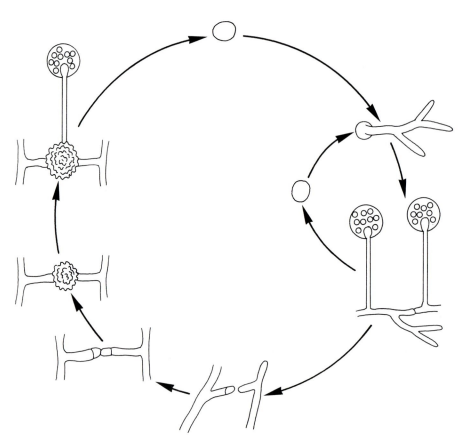

Life Cycle 2: a saprotrophic zygomycete, *Mucor*

The haploid (and monokaryotic) sporangiospore remains dormant until it comes into contact with a suitable substrate, when germination takes place. The haploid mycelium grows through and exploits the substrate. During the process, sporangia are formed, in which there are large numbers of sporangiospores. These are dispersed, again remaining dormant until coming into contact with a suitable substrate. If, and this is somewhat infrequent, two spores with compatible nuclei (see Chapter 11) germinate on the same substrate, sexual reproduction can follow, with the formation of gametangia at the end of short hyphae and, after fusion, zygotes. The latter have thick walls, germinating under favourable conditions to produce a stalked sporangium in which meiosis and many subsequent mitoses occur in the production of spores. The spores are released to start the life cycle again.

(b)

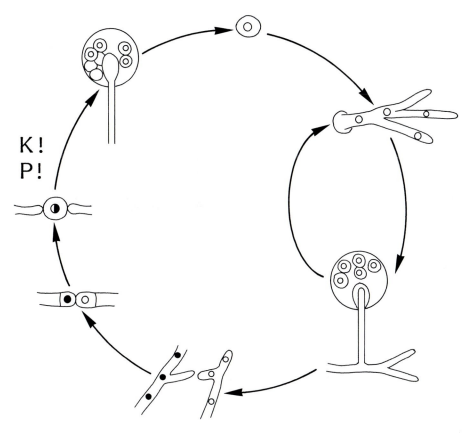

K!
P!

(a)

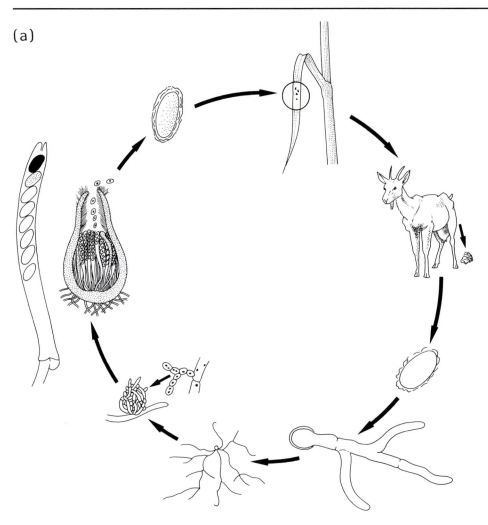

Life Cycle 3: coprophilous ascomycete, *Sordaria humicola*

Monokaryotic ascospores formed in the fruit body (perithecium) are shot on to a grass leaf, where they remain dormant until the leaf is eaten by a grazing animal. As the spore moves through the intestine, its wall becomes softened, such that when dung is deposited, spore germination can readily take place. The resulting mycelium grows through the dung, eventually producing perithecial initials which possess receptive hyphae, or trichogynes, which can absorb (compatible?) nuclei from other mycelia, spermatia or (micro-) conidia. This leads to plasmogamy (P!) followed by the formation of the hook (lower diagram, opposite). The dikaryotic or ascogenous hyphae formed now start to produce asci, in which karyogamy (K!) and meiosis (R!) occur. Before

(b)

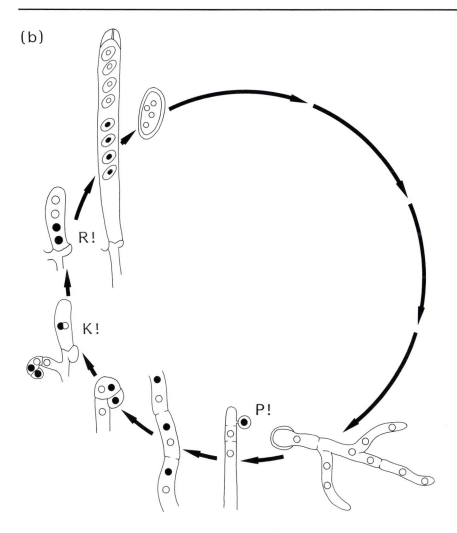

ascospore formation, there is an additional mitosis (M!) which brings the number of spores in the ascus to eight. In certain instances, there can be further mitoses after ascospore formation occurs. In this way, the number of nuclei in the spore can be increased but it is still monokaryotic, since only nuclei of one type are present in the spore.

(a)

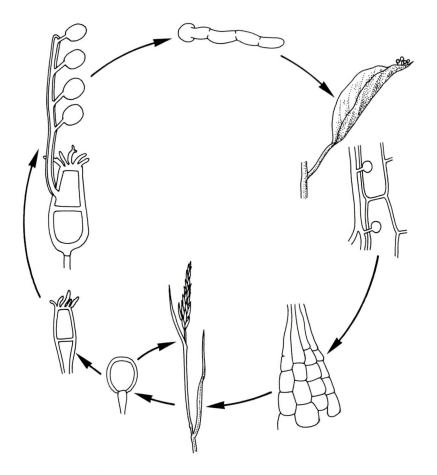

Life Cycle 4: the rust fungus *Puccinia coronata*

The plant parasitic fungus *Puccinia coronata* has two hosts. The monokaryotic basidiospores infect the shrub *Rhamnus*. In this host dikaryotization can occur if there are compatible mycelia. In the cytological diagram, different shapes are used to symbolize those nuclei which are different in terms of genetic incompatibility. Dikaryotic mycelium produces aeciospores which infect the second host, a grass from the genus *Avena*. Following the successful infection of the grass, the mycelium forms uredospores which infect other grass leaves and plants. If this process is repeated, it is not long before there is an epidemic outbreak of the disease amongst the grass

(b)

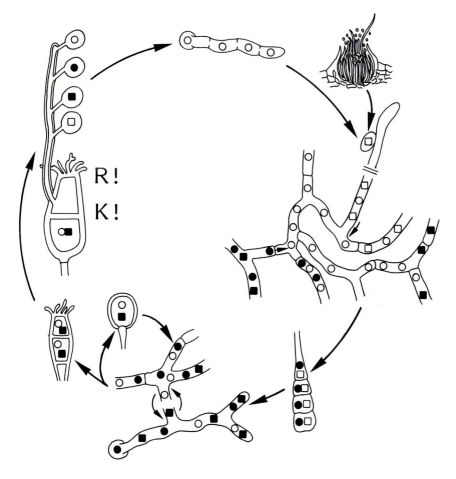

population. At the end of the vegetative period of the *Avena* plant or as the result of exhaustion of the readily available nutrients, teleospores are produced. These spores overwinter. In the following spring, they germinate to produce septated basidia on which monokaryotic basidiospores are produced. These spores restart the life cycle by infecting the first host, *Rhamnus*. However, since infected parts of this shrub remain alive through the winter, the dikaryotic state is maintained during this period, allowing immediate propagation of the fungus without sexual reproduction to new leaves in the following spring.

(a)

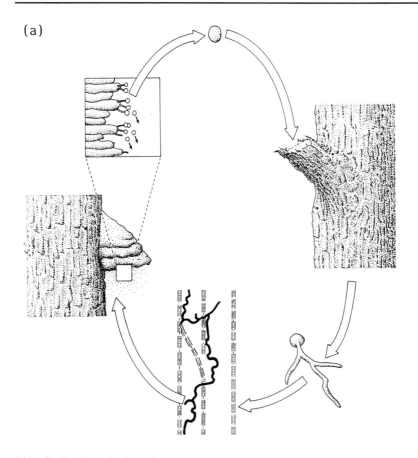

Life Cycle 5: a timber-destroying basidiomycete, a bracket fungus such as *Fomes*

Basidiospores produced by the bracket fruit body reach the wound on a new host, which may have been caused by a branch breaking off. On the wound, the spores germinate and the resulting hyphae extend and come into contact one with another, exchanging compatible nuclei. This results in dikaryotic mycelia, whose nuclei, in the cytological diagram, are symbolized by different shapes according to their genetic type compatibility. The various mycelia extends trophically through the trunk bringing about decay; during this time nuclear exchange continues. After a period, fruit bodies are formed which produce basidia in which karyogamy (K!) and meiosis (R!) occur leading to the production of basidiospores. The lower diagram shows the events leading to the formation of a clamp connection.

(b)

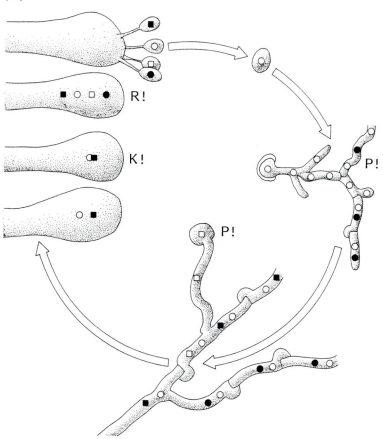

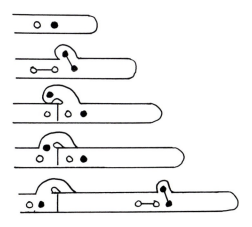

Classification of genera referred to in this book

This is not a complete classification of fungi, nevertheless only a few groups are not represented here. The scheme of classification used here is thought to be the most appropriate for the reader. For other schemes, the reader should consult Hawksworth, D.L., Sutton, B.C. and Ainsworth, G.C. (1983) *Dictionary of Fungi* (7th edn). Commonwealth Institute, Kew.

Kingdom Fungi

Division Eumycota (those fungi in which there is no amoeboid or plasmodial phase and which are typically mycelial but sometimes unicellular).

Class Chytridiomycetes
    Neocallimastix
    Rhizophydium

Class Oomycetes
    Order Peronosporales
        Albugo
        Bremia
        Peronospora
    Order Saprolegniales
        Achlya
        Saprolegnia

Class Zygomycetes
    Order Endogonales
        Endogone
    Order Mucorales
        Choanephora
        Cunninghamella
        Mortierella

    Mucor
    Pilobolus
    Rhizopus
    Syncephalastrum
    Thamnidium
  Order Zoopagales
    Piptocephalis
    Rhopalomyces

Class Ascomycetes (only certain orders are represented)
  Order Clavicipitales
    Atkinsonella
    Balansia
    Claviceps
    Epichlöe
    Myriogenospora
  Order Diaporthales
    Gaeumannomyces
  Order Endomycetales (teleomorphic yeasts)
    Debaryomyces
    Hansenula
    Saccharomyces
    Zygosaccharomyces
  Order Erysiphales
    Erysiphe
  Order Eurotiales
    Eurotium
    Talaromyces
    Thermoascus
  Order Heliotales
    Hymenoscyphus
    Rutstroemia
    Sclerotinia (anamorph: *Monilinia*)
  Order Pezizales
    Monascus
    Morchella
    Saccobolus
  Order Polystigmatales
    Magnaporthe
  Order Rhytismales
    Rhytisma
  Order Sordariales
    Chaetomium
    Humicola
    Neurospora
    Podospora

    Sordaria
  Order Sphaeriales
    Arenariomyces
    Corollospora
    Kohlmeyeriella
    Nereiospora
    Octospora
    Remispora

Class Basidiomycetes
  Order Agaricales
    Agaricus
    Armiellaria
    Clitocybe
    Coprinus
    Marasmius
    Mycaena
    Oudemansiella
    Panaeolus
    Pholiota
  Order Aphyllophorales (Polyporales)
    Coriolus
    Digitatiospora
    Phanerochaete
    Polyporus
    Poria
    Serpula
    Trametes
  Order Boletales
    Leccinum
  Order Melanogastrales
    Nia
  Order Nidulariales
    Cyathus
  Order Phallales
    Clathrus
    Phallus
  Order Uredinales (parasites of plants)
    Puccinia
    Uromyces
  Order Ustilaginales (parasites of plants)
    Ustilago

Class Deuteromycetes (class Fungi Imperfecti - the imperfect fungi)
  Acremonium
  Alternaria

Arthrobotrys
Aspergillus
Aureobasidium
Bispora
Cephalosporium
Chrysosporium
Dactylella
Dendryphiella
Doratomyces
Epicoccum
Fusarium
Harposporium
Monacrosporium
Monilinia
Nematoctonus
Orbimyces
Penicillium
Phoma
Sporobolomyces
Sporotrichum
Trichoderma
Trichospora
Verticillium

Appendix C

Further reading

In this list, we have included details of volumes which superficially might seem out of date but which contain relevant insights as to how fungi behave and function and which are still useful sources of information.

Ahmajdjian, V. and Hale, M.E. (1973) *The Lichens.* Academic Press, London and New York.

Ainsworth, G.C., Sparrow, F.K. and Sussman, A.S. (1973) *The Fungi: an Advanced Treatise* Vol. IVA, *A Taxonomic Review with Keys: Ascomycetes and Fungi Imperfecti*; Vol. IVB, *A Taxonomic Review with Keys: Basidiomycetes and Lower Fungi.* Academic Press, London and New York.

Ainsworth, G.C. and Sussman, A.S. (1965) *The Fungi: an Advanced Treatise,* Vol. 1, *The Fungal Cell;* (1966) Vol. 2, *Fungal Organism;* (1968) Vol. 3, *Fungal Populations.* Academic Press, London and New York.

Alexapoulos, C.J. and Mims, C.W. (1973) *Introductory Mycology.* John Wiley & Sons, Chichester.

Anderson, J.M., Rayner, A.D.M. and Walton, D.W.H. (1984) *Invertebrate–Microbial Interactions.* Cambridge University Press, Cambridge.

Arora, D.K. *et al.* (1991, *et seq.*) *Handbook of Applied Mycology.* Marcel Dekker, New York.

Bainbridge, R.W. (1987) *Genetics of Microbes.* Blackie, Edinburgh.

Beckett, G.C., Heath, I.B. and McLaughlan, D.J. (1974) *An Atlas of Fungal Ultrastructure.* Longman, London.

Bennett, J.W. and Ciegler, A. (1983) *Secondary Metabolism and Differentiation in Fungi.* Marcel Dekker, New York.

Berry, D.R. (1988) *Physiology of Industrial Fungi.* Blackwell Scientific Publications, Oxford.

Berry, D.R., Russell, I. and Stewart, G.C. (1987) *Yeast Biotechnology.* Allen & Unwin, London.

Boddy, L., Marchant, R. and Read, D.J. (1989) *Nitrogen, Phosphorus and Sulphur Utilization by Fungi.* Cambridge University Press, Cambridge.

Brown, A.D. (1990) *Microbial Water Stress Physiology: Principles and Perspectives* John Wiley & Sons, Chichester.

Bünning, E.J. (1973) *The Physiological Clock.* Springer, Berlin.

Burnett, J.H. (1976) *Fundamentals of Mycology.* Edward Arnold, London.

Burnett, J.H. and Trinci, A.P.J. (1979) *Fungal Walls and Hyphal Growth.* Cambridge University Press, Cambridge.

Campbell, R. (1985) *Plant Microbiology.* Edward Arnold, London.

Carlile, M.J. and Watkinson, S.C. (1994) *The Fungi.* Academic Press, New York.

Carrol, G.C. and Wicklow, D.T. (1992) *The Fungal Community: its Organization and Role in the Ecosystem.* Marcel Dekker, New York.

Cochrane, V.W. (1963) *Physiology of Fungi.* John Wiley & Sons, Chichester.

Cole, G.T. and Hoch, H.C. (1991) *The Fungal Spore and Disease Initiation in Plants and Animals.* Plenum Press, New York.

Cooke, R.C. (1977) *Biology of Symbiotic Fungi.* John Wiley & Sons, Chichester.

Cooke, R.C. (1980) *Fungi, Man and his Environment.* Longman, London.

Cooke, R.C. and Rayner, A.D.M. (1984) *Ecology of Saprotrophic Fungi.* Longman, London.

Cooke, R.C. and Whipps, J.H. (1993) *Ecophysiology of Fungi.* Blackwell Scientific Publications, Oxford.

Crawford, R.L. (1981) *Lignin Biodegradation and Transformation.* John Wiley & Sons, Chichester.

Deacon, J.D. (1984) *Introduction to Modern Mycology.* Blackwell Scientific Publications, Oxford.

Dickinson, C.H. and Lucas, J.A. (1982) *Plant Pathology and Plant Pathogens.* Blackwell Scientific Publications, Oxford.

Dix, N. and Webster, J. (1995) *Fungal Ecology.* Chapman & Hall, London.

Elliott, G.C. (1994) *Reproduction in Fungi.* Chapman & Hall, London.

Ellwood, D.C., Hedger, J.N., Latham, M.J., Lynch, J.H. and Slater, J.H. (1980) *Contemporary Microbial Ecology.* Academic Press, New York.

Esser, K. (1986) *Kryptogamen.* Springer, Berlin.

Esser, K. and Kuenen, R. (1967) *Genetics of Fungi.* Springer, Berlin.

Esser, K. and Lemke, P. (1994) *The Mycota,* Vol. I, *Growth, Differentiation and Sexuality* (J.G.H. Wessels and F. Meinhardt, eds). Springer, Berlin. (Six more volumes are promised.)

Fincham, J.R.S., Day, P.R. and Radford, A. (1979) *Fungal Genetics.* Blackwell Scientific Publications, Oxford.

Foster, J.W. (1949) *Chemical Activities of Fungi.* Academic Press, New York.

Frankland, J.C., Hedger, J.N. and Swift, M.J. (1982) *Decomposer Basidiomycetes: their Biology and Ecology.* Cambridge University Press, Cambridge.

Gareth Jones, D. (1983) *Plant Pathology: Principles and Practice.* Open University Press, Stony Stratford.

Gäumann, G. (1952) *The Fungi.* Hafner, New York.

Gow, N.A.R. and Gadd, G.M. (1995) *The Growing Fungus.* Chapman & Hall, London.

Griffin, D.H. (1994) *Fungal Physiology.* Wiley-Liss, New York.

Hale, M.E. (1977) *The Biology of Lichens.* Edward Arnold, London.

Harley, J.L. and Smith, S.E. (1983) *Mycorrhizal Symbiosis.* Academic Press, New York.

Hawksworth, D.L. (1991) *Frontiers of Mycology.* International Mycological Institute, Wallingford.

Hawksworth, D.L. and Hill, D.J. (1984) *The Lichen-Forming Fungi.* Blackie, Edinburgh.

Hawksworth, D.L. and Kirsop, B.E. (1988) *Living Resources for Biotechnology: Filamentous Fungi.* Cambridge University Press, Cambridge.

Hawksworth, D.L., Sutton, B.C. and Ainsworth, G.C. (1983) *Dictionary of the Fungi.* Commonwealth Mycological Institute, Kew.

Ingold, C.T. (1971) *Fungal Spores.* Oxford University Press, Oxford.

Ingold, C.T. and Hudson, H.J. (1993) *The Biology of Fungi.* Chapman & Hall, London.

Isaac, S. (1992) *Fungal–Plant Interactions.* Chapman & Hall, London.

Isaac, S. and Jennings, D.H. (1995) *Microbial Culture.* BIOS Scientific Publishers, Oxford.

Jennings, D.H. (1995) *The Physiology of Fungal Nutrition.* Cambridge University Press, Cambridge.

Jennings, D.H. and Rayner, A.D.M. (1984) *The Ecology and Physiology of the Fungal Mycelium.* Cambridge University Press, Cambridge.

Kendrick, B. (1985) *The Fifth Kingdom.* Mycologue Publications, Waterloo, Ontario.

Kirsop, B.E. and Kurtzman, C.P. (1988) *Living Resources for Biotechnology: Yeasts.* Cambridge University Press, Cambridge.

Large, E.C. (1946) *Advance of the Fungi.* Jonathan Cape, London.

Leong, S.A. and Berka, R.M. (1991) *Molecular Industrial Mycology.* Marcel Dekker, New York.

Manners, J.G. (1993) *Principles of Plant Pathology.* Cambridge University Press, Cambridge.

Michael, E., Hennig, B. and Kreisel, H. (1978 *et seq.*) *Handbuch für Pilzfreunde,* 3rd Edn, 6 Volumes. Gustav Fischer, Berlin.

Moore, D., Casselton, L.A., Wood, D.A. and Rayner, A.D.M. (1985) *Developmental Biology of Higher Fungi.* Cambridge University Press, Cambridge.

Moss, S.T. (1986) *The Biology of Marine Fungi.* Cambridge University Press, Cambridge.

Muller, E. and Loeffler, W. (1971) *Mykologie.* George Thieme, Stuttgart.

Peberdy, J.F., Caten, C.E., Ogden, J.E. and Bennett, J.W. (1991) *Applied Molecular Biology of Fungi.* Cambridge University Press, Cambridge.

Pegg, G.F. and Ayres, P.G. (1990) *Fungal Infection of Plants.* Cambridge University Press, Cambridge.

Ramsbottom, J. (1953) *Mushrooms and Toadstools.* Collins, London.

Rayner, A.D.M. and Boddy, L. (1988) *Fungal Decomposition of Wood.* John Wiley & Sons, Chichester.

Rayner, A.D.M., Brasier, C.M. and Moore, D. (1987) *Evolutionary Biology of the Fungi.* Cambridge University Press, Cambridge.

Robinson, P.M. (1978) *Practical Fungal Physiology.* John Wiley & Sons, Chichester.

Rose, A.H. and Harrison, J.S. (1987) *The Yeasts,* Vol. 1, *Biology of Yeasts*; (1989) Vol. 2, *Yeasts and the Environment*; Vol. 3, *Metabolism and Physiology of Yeasts*; (1991) Vol. 4, *Yeast Organelles*; (1993) Vol. 5, *Yeast Technology.* Academic Press, London.

Ross, I.K. (1979) *Biology of the Fungi.* McGraw-Hill Book Co., New York.

Smith, D.C. and Douglas, A.E. (1987) *The Biology of Symbiosis.* Edward Arnold, London.

Smith, J.E. (1983) *Fungal Differentiation.* Marcel Dekker, New York.

Smith, J.E. and Berry, D.R. (1975) *The Filamentous Fungi,* Vol. I, *Industrial Mycology*; (1976) Vol. II, *Biosynthesis and Metabolism*; (1978) Vol. III, *Developmental Mycology.* Edward Arnold, London.

Smith, J.E., Berry, D.R. and Kristiansen, B. (1983) *The Filamentous Fungi,* Vol. IV, *Fungal Technology.* Edward Arnold, London.

Szaniszlo, P.J. (1985) *Fungal Dimorphism.* Plenum Press, New York.

Wainwright, M. (1992) *An Introduction to Fungal Biotechnology.* John Wiley & Sons, Chichester.

Wicklow, D.T. and Carrol, G.C. (1981) *The Fungal Community.* Marcel Dekker, New York.

Webster, J. (1980) *Introduction to Fungi.* Cambridge University Press, Cambridge.

The following serial publications may also provide relevant review articles:

Advances in Microbial Ecology
Advances in Microbial Physiology
Annual Reviews of Microbiology
Annual Reviews in Phytopathology
Critical Reviews in Biotechnology
Critical Reviews in Microbiology
FEMS Microbiology Reviews
Methods in Microbiology
Microbiological Reviews
Symposia of the Society for General Microbiology
Trends in Biotechnology
Trends in Microbiology

Index